'Haydn Washington shows that everyone has s   the coming
transition. Leadership is critical – each of us mu   to a model and catalyse
inspiration for others, and b) become a well-informed active participant in the political process.'

– William Rees, University of British Columbia, Canada

'Haydn Washington continues to build his series of environmental analyses, here
featuring the theme of healing – from Earth global picture to local "What can I
do?". He challenges us, as he does himself, in the struggle to be honest with ourselves on our landscapes. Confronting overshoot, escalating growth, consumption,
technology, and inequity, we face a broken society, a broken economy, a
wounded nature. At this unprecedented point in world history, we must reinvent
ourselves to re-enchant our wonderland Earth.'

– Holmes Rolston III, University Distinguished Professor,
Colorado State University, USA

'Haydn Washington has done it again! This time he has built upon a spate of previously published and wide-read books on the wonder of nature, our dependence
upon nature, sustainability demystified, steady-state economics, overcoming climate
change denial, and a future beyond growth to write the worried citizen's guide to
restoring a healthy planet. *What Can I Do to Help Heal the Environmental Crisis?*
offers can-do analysis and achievable prescriptions for eco-citizens becoming part of
the solution to our ecological crisis mired in worsening global warming, growing
unsustainable practices, and mounting ecological scarcities. While not for the faint
of heart or for those with their heads in the sand, this work is in many ways the
consummation of this environmental scientist and writer's lifework dedicated to
confronting our environmental predicament, and refusing to give in to cynicism
and despair, His hope is *our* hope – and that of those who come after us.'

– Ernest J. Yanarella, University of Kentucky, USA

'At a time when more people than ever are suffering from solastalgia – mental
distress brought on by seemingly unstoppable environmental change – this book
is an indispensable natural medicine. It turns out that there is much we can all
do to pry open the fingers of the human choke-hold on the natural world.
Washington offers a practical guide to realistic actions everyone can take. And if
everyone does, we might actually heal both our solastalgia and the Earth itself.'

– Robert Engelman, Senior Fellow with the Population Institute and the
Worldwatch Institute, USA

'The human enterprise's unraveling of the tapestry of life is analogous to watch-
ing a train wreck in slow motion. Washington provides a factually solid assessment of changes needed if future generations are to survive. With its cogent
arguments, this book will help those concerned with sustaining life on Earth.'

– Colin L. Soskolne, University of Alberta, Canada

'Haydn Washington's new book forms an excellent segue to his previous vol-
umes reflecting upon the root causes of unsustainability, the need for urgent

reform, both in political/social/economic systems, and in ethics. Washington's book underlines the necessity to heal the world ridden by environmental problems, ranging from biodiversity loss to climate change, and providing excellent ideas of the ways forward.'

    – Helen Kopnina, Hague University of Applied Science, The Netherlands

'How have we come to our current predicament? Is there a way out? Haydn Washington offers clear analysis on how we got here and concrete steps on how each and every one of us can help heal our wounded Earth.'

    – Erik Assadourian, Senior Fellow, Worldwatch Institute, USA

WHAT CAN *I* DO TO HELP HEAL THE ENVIRONMENTAL CRISIS?

The culmination of over three decades of writing by environmental scientist and writer Haydn Washington, this book examines the global environmental crisis and its solutions.

Many of us know that something is wrong with our world, that it is wounded. At the same time, we often don't know why things have gone wrong – or what can be done. Framing the discussion around three central predicaments – the ecological, the social, and the economic – Washington provides background as to why each of these is in crisis and presents steps that individuals can personally take to heal the world. Urging the reader to accept the reality of our problems, he explores practical solutions for change such as the transition to renewable energy, rejection of climate denial and the championing of appropriate technology, as well as a readjustment in ethical approaches. The book also contains 19 'solution boxes' by distinguished environmental scholars.

With a focus on positive, personal solutions, this book is an essential read for students and scholars of environmental science and environmental philosophy, and for all those keen to heal the world and contribute towards a sustainable future.

Haydn Washington is an environmental scientist and writer of over 40 years' experience. He is currently an Adjunct Lecturer at the Pangea Research Centre, School of Biological, Earth and Environmental Sciences, University of New South Wales, Australia. He is the author of seven books on environmental issues, including *Climate Change Denial: Heads in the Sand* (2011 with John Cook), *Human Dependence on Nature* (2013), *Demystifying Sustainability* (2015) and *A Sense of Wonder Towards Nature* (2018).

Routledge Explorations in Environmental Studies

WHAT CAN *I* DO TO HELP HEAL THE ENVIRONMENTAL CRISIS?

Haydn Washington

Routledge
Taylor & Francis Group
LONDON AND NEW YORK

earthscan
from Routledge

First published 2020
by Routledge
2 Park Square, Milton Park, Abingdon, Oxon OX14 4RN

and by Routledge
52 Vanderbilt Avenue, New York, NY 10017

Routledge is an imprint of the Taylor & Francis Group, an informa business

© 2020 Haydn Washington

British Library Cataloguing-in-Publication Data
A catalogue record for this book is available from the British Library

Library of Congress Cataloging-in-Publication Data
A catalog record has been requested for this book

ISBN: 978-0-367-34252-4 (hbk)
ISBN: 978-0-367-34253-1 (pbk)
ISBN: 978-0-429-32466-6 (ebk)

Typeset in Bembo
by Swales & Willis, Exeter, Devon, UK

I dedicate this book to the 'more-than-human' world that has gifted us with so much – and personally taught me the importance of seeking to heal.

CONTENTS

LIST OF FIGURES AND BOXES

Figures

Boxes

BIOGRAPHIES OF BOX AUTHORS

Samuel Alexander is a lecturer and researcher at the University of Melbourne, Australia, teaching a course called 'Consumerism and the Growth Economy: Critical Interdisciplinary Perspectives' as part of the Master of Environment. He is also a research fellow at the Melbourne Sustainable Society Institute and co-director of the Simplicity Institute. Alexander's interdisciplinary research focuses on degrowth, permaculture, voluntary simplicity, 'grassroots' theories of transition and the relationship between culture and political economy. His recent books include *Degrowth in the Suburbs: A Radical Urban Imaginary* (2019, co-authored with Brendan Gleeson) and *Art Against Empire: Toward an Aesthetics of Degrowth* (2017). Most of his work is freely available at www.samuelalexander.info.

Erik Assadourian is a Senior Fellow with the Worldwatch Institute, USA. During his 17 years with Worldwatch, Erik has directed two editions of *Vital Signs* and five editions of *State of the World*, including the 2017 edition: *EarthEd: Rethinking Education on a Changing Planet*, the 2013 edition: *Is Sustainability Still Possible?* and the 2010 edition: *Transforming Cultures: From Consumerism to Sustainability*. Erik also designed *Catan: Oil Springs*, an eco-educational scenario for the popular board game *The Settlers of Catan*. He is also an adjunct professor of Environmental Studies at Goucher College. Erik spends half of each day homeschooling his son.

John Cook is a research assistant professor at the Center for Climate Change Communication at George Mason University, USA. His research focus is understanding and countering misinformation about climate change. In 2007, he founded Skeptical Science, a website which won the 2011 Australian Museum Eureka Prize for the Advancement of Climate Change Knowledge and 2016 Friend of the Planet Award from the National Center for Science Education.

John co-authored the college textbooks *Climate Change: Examining the Facts* and *Climate Change Science: A Modern Synthesis* and the book *Climate Change Denial: Heads in the Sand* (with co-author Haydn Washington).

Brian Czech is the executive director of the Center for the Advancement of the Steady State Economy (www.steadystate.org). Czech served in the headquarters of the U.S. Fish and Wildlife Service from 1999–2017 and as a visiting professor in Virginia Tech's National Capitol Region during most of that period. He is the author of three books: *Supply Shock, Shoveling Fuel for a Runaway Train* and *The Endangered Species Act: History, Conservation Biology, and Public Policy*. Czech has a BS from the University of Wisconsin, an MS from the University of Washington, and a PhD from the University of Arizona, all USA.

Dominick A. DellaSala is President and Chief Scientist of the Geos Institute in Ashland, Oregon, USA, and former President of the Society for Conservation Biology, North America. He is an internationally renowned author of over 200 technical papers and books on forest ecology, endangered species, conservation biology and climate change. Dominick has given keynote talks ranging from academic conferences to the United Nations Earth Summit. He has been featured in hundreds of news stories and documentaries, testified in the US Congress, cofounded organisations and has received conservation leadership and book writing awards. He is on the editorial board of Elsevier, co-editor of the *Encyclopedia of the Anthropocene* and *The World's Biomes*, and editor of scientific journals. Dominick is motivated by his work to leave a living planet for his daughters, grandkids and all that follow.

Robert Engelman is a writer affiliated with the Population Institute and the Worldwatch Institute, USA. From 2014 through 2016 Engelman directed Worldwatch's Family Planning and Environmental Sustainability Assessment project. Earlier he served as President of Worldwatch, an environmental think tank. As a journalist in the late 20th century Engelman reported on science, health and the environment, as well as politics and international affairs, for the Associated Press and numerous daily newspapers. He has written articles for *Scientific American, Nature,* and *Science,* numerous book chapters, and is author of the book *More: Population, Nature, and What Women Want* (2008, Island Press).

Joshua Farley is an ecological economist, Professor in Community Development & Applied Economics, Fellow at the Gund Institute for Ecological Economics at the University of Vermont, USA, and member of the Lab de Sistemas Silvipastoris e Restauração Ecológica (silvopastoral systems and ecological restoration lab) at the Universidade Federal de Santa Catarina, Brazil. His broad research interests focus on the design of economic institutions capable of balancing what is biophysically possible with what is socially, psychologically and ethically desirable. His most recent research focuses on agroecology, farmer livelihoods and ecosystem services in Brazil's Atlantic Forest; the economics of essential resources; and redesigning financial and monetary systems for a just and sustainable economy. He has conducted problem-based courses in ecological economics on six continents. He is co-author with Herman Daly of *Ecological Economics, Principles and Applications*, 2nd edition, from Island Press (2010).

Joe Gray is a field naturalist and eco-activist who lives in the east of England. He has an MSc in Forestry from Bangor University and an MA in Zoology from the University of Cambridge, both UK. Joe is a Fellow of the British Naturalists' Association and the Royal Entomological Society, and in his spare time he serves as county recorder for a group on insects known as terrestrial Heteroptera. He is a co-founder of the recently formed Global Ecocentric Network for Implementing Ecodemocracy (GENIE; https://ecodemocracy.net/). Joe is also a co-founder and associate editor of *The Ecological Citizen* journal and a knowledge adviser on ecological ethics for the United Nations' Harmony with Nature programme.

Stuart B. Hill is Emeritus Professor and Foundation Chair of Social Ecology in the School of Education at Western Sydney University, Australia. He has published over 350 papers and reports. His latest books are *Ecological Pioneers: A Social History of Australian Ecological Thought and Action* (with Dr Martin Mulligan; Cambridge UP, 2001), *Learning for Sustainable Living: Psychology of Ecological Transformation* (with Dr Werner Sattmann-Frese; Lulu, 2008) and *Social Ecology: Applying Ecological Understanding to our Lives and our Planet* (with Dr David Wright and Dr Catherine Camden-Pratt; Hawthorn, 2011). Stuart recently received a Lifetime Achievement 'Leadership for Sustainability Award' from Australia's Centre for Sustainability Leadership.

Helen Kopnina (PhD Cambridge University, 2002) is currently employed at The Hague University of Applied Science (HHS) in The Netherlands, coordinating the Sustainable Business programme and conducting research within three main areas: environmental sustainability, environmental education and biological conservation. Helen is the author of over 90 peer-reviewed articles and (co) author and (co)editor of 15 books.

Michelle Maloney is a lawyer who specialises in Earth jurisprudence and Earth-centred law and governance. She is the Co-Founder and National Convenor of the Australian Earth Laws Alliance, and designs and manages AELA programmes and events, including the Australian Peoples' Tribunal for Community and Nature's Rights. Michelle is the Co-Founder and Director of the New Economy Network Australia (NENA) and a Senior Adjunct Fellow at the Law Futures Centre, Griffith University, Australia. She is also on the Executive Committee of the Global Alliance for the Rights of Nature (GARN) and the Steering Group for the International Ecological Law and Governance Association (ELGA).

Vance G. Martin is an expert in international nature conservation and wilderness protection. Martin specialises in bridging the interests of people and nature through culture, communications, science and policy. President of the WILD Foundation since 1983 – an international organisation dedicated to protecting wilderness, connecting wild nature to people and to communicating the values of wilderness and its benefits for human communities – he has worked in 80+ nations, lived in many countries and founded/served on the boards of numerous organizations. He is also founder and Co-Chair of the Wilderness Specialist Group of the World Commission on Protected Areas (WCPA/IUCN).

Kathleen Dean Moore is the former Distinguished Professor of Philosophy at Oregon State University, USA. She now writes and speaks full time about the moral urgency of climate action. She is the author of *Great Tide Rising: Toward Courage and Moral Clarity in a Time of Planetary Change* and co-editor of *Moral Ground: Ethical Action for a Planet in Peril* and *Witness: The Human-Rights Impacts of Fracking and Climate Change* (forthcoming). Her newest work is *The Extinction Variations (Meadowlarks)*, a music/spoken-word collaboration with concert pianist Rachelle McCabe, performed in the United States, UK and Canada. See: www.riverwalking.com and www.musicandclimateaction.com.

Michael Paul Nelson holds the Ruth H. Spaniol Chair of Renewable Resources and is Professor of Environmental Ethics and Philosophy at Oregon State University, USA. He also serves as the Lead Principal Investigator for the HJ Andrews Long-Term Ecological Research Program in Oregon's Cascade Mountains and as the Philosopher in Residence for the Isle Royale Wolf-Moose Project. An intensely interdisciplinary scholar, he is the author and editor of many articles and books at the interface of environmental ethics, ecology, conservation and social science, including *Moral Ground: Ethical Action for a Planet in Peril*, an award-winning book co-edited with Kathleen Dean Moore. You can see more of Michael's work at www.michaelpnelson.com.

Kate Pickett, FRSA, is a British epidemiologist who is Professor of Epidemiology in the Department of Health Sciences at the University of York, UK, and was a National Institute for Health Research Career Scientist from 2007–12. She co-authored (with Richard G. Wilkinson) *The Spirit Level: Why More Equal Societies Almost Always Do Better* and is a co-founder of The Equality Trust. Pickett was awarded a 2013 Silver Rose Award from Solidar for championing equality and the 2014 Charles Cully Memorial Medal by the Irish Cancer Society.

Holmes Rolston III, is University Distinguished Professor and Professor of Philosophy, Emeritus at Colorado State University, USA. He has written: *Three Big Bangs: Matter-Energy, Life, Mind; Genes, Genesis and God; Philosophy Gone Wild; Environmental Ethics; Science and Religion: A Critical Survey*; and *Conserving Natural Value*. Rolston was laureate for the 2003 Templeton Prize in Religion and gave the Gifford Lectures at Edinburgh in 1997–8. He has spoken as distinguished lecturer on all seven continents, and is featured in Joy A. Palmer Cooper and David E. Cooper, eds., *Key Thinkers on the Environment*, 2018. See: https://en.wikipedia.org/wiki/Holmes_Rolston_III and https://mountainscholar.org/handle/10217/100484.

Colin L. Soskolne, PhD, hails from South Africa. His award-wining PhD is from the University of Pennsylvania, USA. With over 350 peer-reviewed publications, his academic career was one of innovative teaching and research focusing on occupational, environmental and global health. After 28 years as Professor of Epidemiology at the University of Alberta, Canada, he retired as Professor emeritus in 2013. Among other contributions, he led WHO's 1999 *Global Ecological Integrity and 'Sustainable Development': Cornerstones of Public Health*, edited *Sustaining Life on Earth: Environmental and Human Health through Global Governance*, and established the academic arm of the University's Office of Sustainability (www.colinsoskolne.com).

David Tacey is Emeritus Professor of Literature, based in Melbourne, Australia, and author of several books on spirituality and global issues, including *Re-enchantment* (Sydney: Harper Collins 2000), *The Spirituality Revolution* (London: Routledge, 2004) and *The Postsecular Sacred* (London: Routledge, 2019).

Peter A. Victor, author of *Managing without Growth: Slower by Design, not Disaster*, is a Professor Emeritus at York University, UK. He has worked for nearly 50 years in Canada and abroad on economy and environment issues as an academic, consultant and public servant. His work on ecological economics has been recognised through the award of the Molson Prize in the Social Sciences by the Canada Council for the Arts in 2011, the Boulding Memorial Prize from the International Society for Ecological Economics in 2014, and election to the Royal Society of Canada in 2015.

Richard G. Wilkinson is a British social epidemiologist, author and advocate. He is Professor Emeritus of Social Epidemiology at the University of Nottingham, UK, having retired in 2008. He is also Honorary Professor of Epidemiology and Public Health at University College London and Visiting Professor at University of York, UK. In 2009, Richard co-founded The Equality Trust. Richard was awarded a 2013 Silver Rose Award from Solidar for championing equality and the 2014 Charles Cully Memorial Medal by the Irish Cancer Society.

Ernest J. Yanarella is a Professor and former Department Chair in the Political Science Department at the University of Kentucky, USA. For several decades, he served as the Associate Director of the Center for Sustainable Cities and Director of Environmental Studies Program. His primary teaching and research interests are in the areas of critical policy studies (urban, energy and environment, and agricultural and ecological policy), political theory, sustainable urban design, and politics and literature. His recent books are *The City as Fulcrum of Global Sustainability* (2011) and *Getting from Here to There? Power, Politics, and Urban Sustainability in North America* (2016).

FOREWORD

'Scorched Earth'

William E. Rees, PhD, FRSC

UNIVERSITY OF BRITISH COLUMBIA, ORIGINATOR AND CO-DEVELOPER OF THE ECOLOGICAL FOOTPRINT CONCEPT

This book was compelled by truly extraordinary trends occurring on Earth today. Those who track global change will know that atmospheric carbon dioxide (CO_2) readings in April 2018 averaged over 410 parts per million (ppm) for the entire month for the first time on record (NOAA 2018). Just a year later, in early May 2019, CO_2 concentrations exceeded 415 ppm for the first time (NOAA 2019). Human activities have elevated atmospheric carbon to its highest level in 800,000 years, 46% above the pre-industrial baseline of 280 ppm.

Carbon dioxide is an inevitable entropic by-product of fossil fuel combustion and the greatest metabolic waste by weight of industrial economies. It also happens to be the main anthropogenic driver of climate change. No one should be surprised, therefore, that mean global temperature has climbed by approximately 1°C above the pre-industrial mark, mostly since 1980. The first 18 years of this century boast a statistically improbable 17 of the 18 warmest years since record-keeping began and 'the past five years are, collectively, the warmest years in the modern record' (NASA 2019). When the world was last this warm (the Eemian period 130,000–115,000 years ago) sea levels eventually rose 6–9 metres (20–30 feet), sufficient to drown most of today's coastal plains, towns and cities and displace tens of millions of people.

Ominously, climate science tells us that the world is on track for 3–5°C warming. A 5°C uptick would be catastrophic, likely fatal to civilization. Even a 'modest' 3°C implies disaster – enough to destroy economies, empty many cities and destabilize geopolitics. The 2015 COP 21 Paris Agreement therefore 'committed' Parties to the United Nations Framework Convention on Climate Change to hold the rise in global average temperatures to 'well below 2°C above pre-industrial levels and pursue efforts to limit the temperature increase to 1.5°C above pre-industrial levels'. Sadly, climate action typically falls short of aspiration. Even if fully met, the actual emissions reduction *pledges* made in Paris would lead to >3°C warming.

And there is yet another problem – recent analysis suggests that:

> [b]iogeophysical feedback processes [e.g., permafrost melting, methane hydrate releases, tropical and boreal forest die-back] within the Earth System ... play a more important role than normally assumed ... there is a significant risk that these internal dynamics, especially strong nonlinearities in feedback processes, could become an important or perhaps, even dominant factor.
>
> *(Steffen et al. 2018: 2)*

Indeed, 'even if the Paris Accord target of a 1.5 C to 2.0 °C rise in temperature is met, we cannot exclude the risk that a cascade of feedbacks could push the Earth System irreversibly onto a "Hothouse Earth" pathway' (Steffen et al. 2018: 3). It complicates matters that a ~40-year lag between cause and effect means that even if atmospheric GHG concentrations remain *constant*, the world will experience an additional 0.3–0.8°C degree warming in coming decades, *enough to overshoot the 1.5 degree limit* (Hansen 2018). Meanwhile, CO_2 levels are actually *climbing* by 3 ppm annually and other greenhouse gas concentrations are rising as fast or faster.

Rockström et al. (2017) argue that, to neutralize climate risk, the world community must cut fossil fuel use in half each decade until 2050 as well as extract gigatonnes (Gt) of carbon out of the atmosphere all while shifting society to alternative green energy. Similarly, the IPCC Special Report on 1.5°C warming and possible emissions pathways found that in 'model pathways with no or limited overshoot of 1.5°C, global net anthropogenic CO_2 emissions decline by about 45% from 2010 levels by 2030 ... reaching net zero around 2050' (IPCC 2018, C1). All such pathways require 'the use of carbon dioxide removal (CDR) on the order of 100–1000 Gt CO_2 over the 21st century' (IPCC 2018: C3). No one is sure what form a CDR program might take.

Sadly, even beating back climate change wouldn't put us – or the ecosphere – out of danger. As Haydn Washington details in the chapters that follow, other existential problems confound the modern human 'predicament'. Consider humanity's role in biodiversity loss. Few people are fully aware that *H. sapiens* competes directly and indirectly with other vertebrates for Earth's finite liveable space and food energy – and, thanks to technology, we are literally eliminating the competition. With only 0.01% of total earthly biomass, *H. sapiens* has 'competitively displaced' 83% of wild animal and removed 50% of natural plant biomass. From a fraction of 1% ten millennia ago, humans now constitute 36%, and our domestic livestock another 60%, of the planet's mammalian biomass compared to only 4% for all wild mammals combined. Similarly, domestic poultry now comprises 70% of Earth's remaining avian biomass (Bar-On et al. 2018). Meanwhile, the oceans are being depleted by commercial fishing –

which is itself one cause of rapidly declining marine mammal and sea-bird populations. 'Indeed, seabirds are the most threatened bird group, with a 70% community-level population decline across 1950–2010' (Grémillet et al. 2018). Overall, the World Wildlife Fund reports an 'astonishing' 60% decline in the populations of mammals, birds, fish, reptiles, and amphibians in just over 40 years (WWF 2018). Other plant and animal groups are also under siege from habitat loss, insecticides, industrial contamination and, of course, climate change – even insects are experiencing 'Armageddon' and systemically-linked insect-dependent birds, mammals and amphibians are not far behind (Hallmann et al. 2017; Lister and Garcia 2018).

Few of the world's increasingly urban citizens are conscious of the accelerating pace of change. It took a laggardly 200,000 years for the human population to reach one billion in the early 19th century but just the next 200 years – 1/1000th as much time – to top 7.7 billion by 2018. Meanwhile, between 1815 and 2015, real global GDP increased about 100-fold and average per capita incomes by a factor of 13 (rising to 25-fold in the richest countries) (Roser 2018). Material consumption and attendant pollution kept pace with income growth, driving the runaway degradation of air, land, water and ecosystems (see Steffen et al. 2015). Bottom line? During the 20th century, our species became not only the most destructive ecological entity ever but also the major geological force altering the face of Earth. (It is more than a little ironic that global desecration is the clearest mark of humanity's evolutionary 'success'.)

Nor do most people realize that explosive techno-industrialization has mostly been a fossil-fuelled phenomenon – abundant cheap energy provides the means for *H. sapiens* to plunder the ecosphere for the food and other material resources required to expand the human enterprise. Exponential accelerator pressed to the floor, humans have consumed half the fossil energy ever burned (and spewed half of the fossil CO_2 ever emitted), in just the past 25–30 years!

These data make clear that our contemporary environmental 'predicament' is the product of *gross human ecological dysfunction*. The human enterprise is in plague-like overshoot, consuming food, fibre and other bio-resources faster than ecosystems can regenerate and filling natural waste sinks beyond nature's assimilative capacity (from this perspective, climate change is a waste management problem!). *H. sapiens* is quite literally converting the natural world into more human bodies, cultural artifacts and supportive infrastructure; we are financing material growth by liquidating species and ecosystems essential for our own survival, all while polluting what remains of our increasingly tattered nest.

It is all the more remarkable then that mainstream governments and development organizations see the future as a technologically more advanced and expansive version of the recent past, i.e. as a recharged 'business-as-usual'. Modern humans remain in thrall of their own socially-constructed progress myth, oblivious that the two centuries of growth they take to be the norm, actually define the most *anomalous* period in 200 millennia of human history.

They also seem unaware that what goes up must inevitably come down. The human enterprise can be maintained in a highly organized, smoothly functioning far-from-(thermodynamic)-equilibrium state *only* through the continuous consumption of prodigious quantities of energy, currently mostly fossil fuels. And herein lies a conundrum. Continued reliance on fossil energy implies climate chaos and the destruction of civilization. Without suitable substitutes for fossil fuel, civilization can only implode and suffocate in its own entropic excreta. And, at present, despite much popular hype, there are no fully satisfactory green alternatives for coal, oil and natural gas – wind and solar photovoltaics simply don't cut it. One other thing – even if there were equivalent substitutes, *without a transformation of values and economic behaviour*, humans would simply use the new energy to continue their destructive dissipation of the ecosphere.

Extraordinary circumstances demand extraordinary responses. The only more or less satisfactory resolution of our predicament is a planned economic and population contraction, perhaps towards a global system of smaller, more equitable steady-state regional economies each functioning within the regenerative means of nature. If each quasi-independent entity achieves regional sustainability, the whole will be sustainable.

Achieving a global 'soft-landing' is clearly a collective enterprise requiring collective solutions – no individual or nation can 'go it alone'. However, as Haydn Washington elaborates below, every person has many vital roles to play in the coming transition. In essence, he shows that each of us must strive to: a) be a catalytic model and inspiration for others to follow and b) become a well-informed active participant in the political process. (If you don't become an irritating thorn in the side of every intransigent politician, you won't be doing your job.)

People today must exploit every opportunity for progressive policy change as if their lives depend on it – because they do. Let us move past denial and accept the grim reality of our predicament. Only then can we empower ourselves to implement the positive (if challenging) solutions that exist to heal our world.

References

Bar-On, Y. M., Phillips, R. and Milo, R. (2018) 'The biomass distribution on earth', *National Academy of Sciences*, 21 May, 2018 201711842. see: www.pnas.org/content/early/2018/05/15/1711842115 (accessed 15 May 2019).

Grémillet, D., Ponchon, A., Paleczny, M. et al. (2018) 'Persisting worldwide seabird–fishery competition despite seabird community decline', *Current Biology*, 28 (24): 4009–4013. https://doi.org/10.1016/j.cub.2018.10.051.

Hallmann, C. A., Sorg, M., Jongejans, E. et al. (2017) 'More than 75 percent decline over 27 years in total flying insect biomass in protected areas', *PLoS One*, 12 (10): e0185809. https://doi.org/10.1371/journal.pone.0185809.

Hansen, J. (2018) 'Climate change in a nutshell: The gathering storm', see: www.columbia.edu/~jeh1/mailings/2018/20181206_Nutshell.pdf (accessed 10 May 2019).

IPCC. (2018) *Global Warming of 1.5°C – Summary for Policymakers*, Intergovernmental Panel on Climate Change, see: www.ipcc.ch/sr15/chapter/summary-for-policy-makers/ (accessed 2 December 2018).

Lister, B. C. and Garcia, A. (2018) 'Climate-driven declines in arthropod abundance restructure a rainforest food web', *PNAS*, 115 (44): E10397–E10406, see: www.pnas.org/content/115/44/E10397 (accessed 10 May 2019).

NASA. (2019) '2018 fourth warmest year in continued warming trend, according to NASA, NOAA', National Aeronautics and Space Administration, see: https://climate.nasa.gov/news/2841/2018-fourth-warmest-year-in-continued-warming-trend-according-to-nasa-noaa/ (accessed 15 May 2019).

NOAA. (2018) 'Trends in atmospheric carbon dioxide', National Oceanic and Atmospheric Administration, see: www.esrl.noaa.gov/gmd/ccgg/trends/ (accessed 25 August 2018 (updated frequently)).

NOAA. (2019) 'Trends in atmospheric carbon dioxide', National Oceanic and Atmospheric Administration, see: www.esrl.noaa.gov/gmd/ccgg/trends/weekly.html (accessed 20 May 2019 (updated frequently)).

Rockström, J., Gaffney, O., Rogelj, J. et al. (2017) 'A roadmap for rapid decarbonisation', *Science*, 355 (6331): 1269–1271. 10.1126/science.aah3443.

Roser, M. (2018) 'Economic growth', Our World in Data, see: https://ourworldindata.org/economic-growth (accessed 10 May 2019).

Steffen, W., Broadgate, W., Deutsch, L. et al. (2015) 'The trajectory of the anthropocene: The great acceleration', *The Anthropocene Review*, 2 (1): 81–98.

Steffen, W., Rockström, J., Richardson, K. et al. (2018) 'Trajectories of the earth system in the anthropocene', *PNAS*, 115 (33): 8252–8259. https://doi.org/10.1073/pnas.1810141115.

WWF. (2018) *Living Planet Report*, Gland, Switzerland: Worldwide Fund for Nature.

ACKNOWLEDGMENTS

I would like to thank those who made comments on the draft book, especially the (self-identified) reviewers Helen Kopnina of the Hague University of Applied Science, Ernest Yanarella of the University of Kentucky and Erik Assadourian of the Worldwatch Institute. I would also like to thank my other reviewer who remained anonymous. I would like to thank those that made comments on particular chapters, especially Stuart Hill of Western Sydney University, Robert Engelman of the World-watch Institute and Michelle Maloney of the Australian Earth Laws Alliance. I would also like to thank the many box authors who took the time to respond to the invitation to write a box for the book. Their involvement has added an extra and special dimension to the book. I would like to thank Ernest Yanarella for the idea of adding boxes by various scholars. I would like to thank Erik Assadourian for the idea of having educational exercises at the end of the book. Of course, I would also like to thank my contacts at Routledge, Annabelle Harris and Matthew Shobbrook, for their support for the project.

I would also like to thank the nonhuman persons who have inspired me to write about healing the planet. I live on the edge of Wollemi National Park in NSW, and this was the living land I had key involvement with as a conservationist in creating it as a National Park in the 1970s. It is also my *spiritual home*. My favourite phrase is 'If you listen you will learn' and this land has taught me so much! And I am still listening ... and still learning.

(Cover photo: The cover photo for the paperback edition is of a Yellow Box tree (*Eucalyptus melliodora*) on the author's property in Australia that is several hundred years old. Decades ago an attempt was made to ring-bark the tree to promote the growth of grass for grazing. However, the tree has healed itself by growing over the ring cut around its circumference. It is thus a fitting image for healing the environmental crisis. Photo by the author)

INTRODUCTION

It must not be
Sustain-a-babble
Nor denial, nor tokenism.
Its heart should be
A loving engagement,
Caring commitment
To our greatest work
In four parts:
Listen to the land
Respect the land
Celebrate the land
Heal the land.

From 'Sustainability' (Washington 2013a)

What can anyone do to help heal the environmental crisis, or indeed to heal the world? I can understand that some readers might think this is an arrogant question, maybe even one that shows hubris? However, I believe this question is one we should ask ourselves at least every week, probably every day. Asking such a question clearly assumes that we need to heal the world. Chapter 1 is a summary of the facts, which show beyond doubt that the world *is* wounded (the Foreword also shows this clearly). I will state right at the start that I assume throughout this book that the purpose of life is not to make money, to gain status or win power. Rather, the purpose of our lives should be to do *something of value*, and there is nothing intrinsically valuable about a large bank balance, nor a fight for status and power. Rather, many wise people over history have argued that we should seek to help others, to heal them. Here, I just expand the ethics of this to include not just human 'persons' but also nonhuman persons (Harvey 2005) and the whole world.

The scientific data is well and truly in; quite simply we are trashing our world (e.g. Brown 2011; Wijkman and Rockstrom 2012; AWS 2017). There are many reasons behind this, and this book explores these. However, I think any thoughtful person can understand our predicament. I use the term 'predicament' here, which has a meaning of: 'a difficult, unpleasant or embarrassing situation'. Synonyms for predicament rightly include 'quandary', 'muddle' and 'mess' (all of these are valid for our situation). Indeed, humanity is in a very difficult situation, as our actions are degrading the natural world that supports our society. The situation is clearly becoming increasingly unpleasant, as the discussion later explains. We should indeed be embarrassed that we have let it get to this stage, where we have an escalating crisis because society denies both its problems *and* their solutions.

In regard to our predicament, just look around you and do a bit of research. The human population is expanding, pushing out the nonhuman from their homes everywhere, causing mass extinction and decline of ecosystems – 'ecocide' (Wijkman and Rockstrom 2012; Washington 2013b; Crist et al. 2017). Species are going extinct (never to return), forests are falling, ecosystems collapsing, pollution is increasing and masses of plastics desecrate the oceans, killing many marine animals. Indeed theologian Thomas Berry (1988: 7) succinctly states Western society's key problem:

> We can break the mountains apart; we can drain the rivers and flood the valleys. … We can pollute the air with acids, the rivers with sewage, the seas with oil – all this in a kind of intoxication with our power for devastation … And why? To increase the volume and speed with which we move natural resources through the consumer economy to the junk pile or the waste heap. … If the environment is made inhospitable for a multitude of living species, then so be it. … But our supposed progress toward an ever-improving human situation is bringing us to a wasteworld instead of a wonderworld.

Yet people are not made 'happier' by this growth fixation, it does not expand our wisdom or help us become better people. Quite the reverse. Society is not happier; indeed inequality is increasing, and trust and social capital are on the decline. Everywhere a keen observer may care to look – we see the world is wounded. So we should ask, do I wish to assist this malaise *or* seek to heal it?

Now one could trot out various trite sayings as to why you or I cannot possibly 'make a difference'. The first is 'You cannot fight city hall', meaning you cannot successfully oppose bad government. However, this is clearly mistaken. Most social reforms have happened because people lobbied their governments successfully to force change. That is how change happens. Whether it was banning slavery, seeking universal suffrage so women too could vote, or gaining environmental laws to limit environmental degradation – these things have always come from people lobbying for what is right. So you *can* fight city halls – and win – and many people have (myself included). Another common statement is: 'It is too late!' This of course actually becomes a self-fulfilling prophecy, for if one does nothing, clearly nothing will change. I am an environmental scientist and well aware of the seriousness of the environmental and climate crises. Indeed, I have at times had to face depression about what is happening. We are indeed skating on thin ice in terms of passing 'points of no return'. Examples include moving close to tipping points for runaway climate change (Spratt and Dunlop 2018) and for ecosystem collapse (see Chapter 10). However, doing nothing to change things just ensures we will indeed pass such points. Taking action means we may just be able to avoid this. Despair is like denial in one key aspect – you do nothing. On the other hand, blind optimism not based on reality (as discussed in Chapter 10) is also not a positive strategy.

To put the need for healing in perspective, despite the depressing news in Chapter 1 in regard to the state of the world, I would like to point out that the world is *wounded not dying*. It has taken me over 40 years to fully understand this distinction. Make no mistake, the extinction of species, the collapse of ecosystems and the ecocide underway (leading towards the collapse of societies) *is* a tragedy, one we should rightly grieve over. We should understand that we may (without action) send half of life on Earth extinct (Wilson 2003) – including perhaps ourselves. However, the creative principle of nature will still go on and evolve past such tragedies. That in no way justifies what society is doing (which can, I believe, be correctly described as a great moral crime). However, it does tell us that in no way is it 'all over' for life and nature. While nature and the Universe exist, it will never be 'all over', as the 'Will to Live' (Schopenhauer 1983) will act to evolve more life. Hence there is no excuse for the statement 'It is too late!' as discussed in Chapter 10. It is not too late – and never will be. So the world is wounded not dying, and it is up to us to help heal that wound. This healing will not be finished in our lifetimes; it will indeed be 'the long emergency' (Orr 2013). However, our personal actions for such healing do two things. Firstly, they are part of the many small acts that add up to healing the world (Hill 1998). Secondly, they are a focus that gives our lives meaning and moral integrity (see Chapter 10). Both are acts of great value.

It is also crucial to realise that we can make a difference. *Why then don't we?* This book discusses several causes that underlie society's lack of action. These range from a human tendency to deny unpleasant realities; to a dominant and destructive anthropocentric worldview; to apathy; to commitment to various dangerous 'isms' and assumptions; to an ecologically unsustainable economics; and to simple arrogance and hubris. These crowd our lives, hedging us in, deluding us as to both the key problems *and* their solutions (Washington 2015, 2018). Some of us are so deluded that we may even think things are 'going well', that humanity is heading towards an exciting future (e.g. Porritt 2013), not towards collapse (see Chapter 10). Hence some in society may be sublimely indifferent to the current grim situation we face. Others will angrily deny any problems (see Chapter 3). Others will see the problems and despair as they cannot see solutions, leading to apathy, a feeling of powerlessness and a failure to take action. Yet – we can indeed act.

This book will look at the problems and barriers to healing our world, and seek ways past them. The science tells us very clearly that there is no time left for denial and despair. It is time to *act*, to cut through the waffle and force change. There are solutions, some fairly easy, some much harder. We will (as a society and individually) need to transform, to embrace positive change based on accepting our problems and solving them. It will not be easy or fun, it won't offer us 'more' physically than we have now. Indeed, most of us will need to make sacrifices, to abandon seeking to be (in the West) princes and princesses. Rather, we will need to become a people in harmony with the rest of life (Washington 2018), being eco-citizens not 'Masters of the Universe'. We never actually were 'masters' of anything, and this fantasy is well and truly past its use-by date. Transformation may make us better, more caring, people who live sustainably, but it won't make us richer people financially, with more toys and techno-junk. We will not travel at warp speed as in *Star Trek* around a Universe we dominate – and egotistically imagine we 'own'. We never 'owned' it, just as we can never philosophically or ethically own the Earth or the land around us (see Chapter 8). This was just another fantasy we created. Now don't get me wrong, I *like* science fiction and fantasy, but I do realise one is fiction and the other fantasy. We need now to face two key things. The first is *facts*, the facts of the destructive predicament we are in. The second is *ethics*, which is about right and wrong. Is it right what we are doing to ourselves, other people, future generations and the wondrous life we are lucky enough to share this world with? Or is it immensely wrong? Is it time for society to accept that what it is doing is a great moral crime at so many levels? This book explains why my answer is emphatically 'Yes!' As we will find out through the book, in a world often lacking in meaning, seeking to heal gives us a deep positive purpose. It provides meaning to our lives.

That is why we should seek to heal our world. And there are many things we can do right now. We are not powerless; we just need to *act*.

References

AWS. (2017) 'World scientists' warning to humanity: A second notice', Alliance of World Scientists, see: http://scientistswarning.forestry.oregonstate.edu/sites/sw/files/Warning_article_with_supp_11-13-17.pdf (accessed 2 March 2019).

Berry, T. (1988) *The Dream of the Earth*, San Francisco: Sierra Club Books.

Brown, L. (2011) *World on the Edge: How to Prevent Environmental and Economic Collapse*, New York: W.W. Norton and Co.

Crist, E., Mora, C. and Engelman, R. (2017) 'The interaction of human population, food production, and biodiversity protection', *Science*, **356**: 260–264.

Harvey, G. (2005) *Animism: Respecting the Living World*, London: Hurst and Co.

Hill, S. B. (1998) 'Redesigning agroecosystems for environmental sustainability: A deep systems approach', *Systems Research and Behavioural Science*, **15**: 391–402.

Orr, D. (2013) 'Governance in the long emergency', in *State of the World 2013: Is Sustainability Still Possible?*, ed L. Starke, Washington: Island Press, pp. 279–291.

Porritt, J. (2013) *The World We Made*, London: Phaidon Press.

Schopenhauer, A. (1983). *The Will to Live: The Selected Writings of Arthur Schopenhauer*, ed R. Taylor, New York: Ungar.

Spratt, D. and Dunlop, I. (2018) 'What lies beneath: the scientific understatement of climate risks', Melbourne, Australia: Breakthrough - National Centre for Climate Restoration. see: https://climateextremes.org.au/wp-content/uploads/2018/08/What-Lies-Beneath-V3-LR-Blank5b15d.pdf.

Washington, H. (2013a) *Poems from the Centre of the World*, lulu.com, see: www.lulu.com/shop/haydn-washington/poems-from-the-centre-of-the-world/paperback/product-21255751.html (accessed 27 March 2019).

Washington, H. (2013b) *Human Dependence on Nature: How to Help Solve the Environmental Crisis*, London: Earthscan.

Washington, H. (2015) *Demystifying Sustainability: Towards Real Solutions*, London: Routledge.

Washington, H. (2018) *A Sense of Wonder Towards Nature: Healing the World through Belonging*, London: Routledge.

Wijkman, A. and Rockstrom, J. (2012) *Bankrupting Nature: Denying Our Planetary Boundaries*, London: Routledge.

Wilson, E. O. (2003) *The Future of Life*, New York: Vintage Books.

1

OUR PREDICAMENT

Why we need to heal the world

It is not easy
To witness
Tragedy in motion,
To stand and watch
The roller coaster
Rush down above the chasm,
Where the track ends …

From 'Mind Trap' (Washington 2019a)

SUMMARY

Human society is completely dependent on the ecosystems that supply us with the services and benefits we need, yet we are degrading these in a major way. The chapter argues that our society is broken, our economy is broken and the nature that supports us is breaking. It considers each one, starting first with our ecological predicament, where we are in overshoot of sustainable ecological limits. Environmental indicators universally show an environmental crisis. The Global Ecological Footprint is 1.7 Earths (yet we have only one). The Living Planet Index has dropped 60% since 1970. A major biodiversity extinction event is underway, where (without change) half of life on Earth may be extinct by 2100, and 60% of the ecosystem services that support society are degrading or being used unsustainably. We have exceeded at least four of the nine suggested 'Planetary Boundaries'. The chapter also considers our social predicament, where equity and equality are declining, as is social capital. The chapter finishes by looking briefly (as it is covered further in Chapters 3 and 5) at our economic predicament. The key problem here is our addiction to an 'endless growth' economy – when we live on a finite planet that society has forced past ecological limits. These three predicaments comprise the key aspects of our overall predicament, where the world needs major and rapid healing. For the detail explaining this summary, please read on; however, if you just want to know the things *you can do* (21 are listed) about this, then turn to the end of the chapter.

Introduction

I should warn the reader up front, do not read this chapter unless you are *also* going to read Chapter 11 on solutions. Otherwise one can be overwhelmed by the problems without realising that positive and practical solutions do exist.

So is our predicament so serious that we really do need to heal the world? In my book *Demystifying Sustainability* (Washington 2015: 195), I claim that: 'our economy is broken, our society is broken, and that the ecosystems that support us are breaking'. Why would I say such a thing? I discuss the reasons here. The first thing to do when considering our predicament is to discuss *ecological reality*. Like all species, humans are fully dependent on their ecosystems to survive (Cardinale et al. 2012; Washington 2013). Humanity of course also has culture and technology, and these allow us to consume more than we could have otherwise (see Chapter 6). However culture and technology enable us to expand the human use of ecosystems, it doesn't change the reality that, in the end, society has obligate dependence on nature and must (like all species) *live within ecological limits*. I will summarise these three key predicaments – 1) The ecological predicament; 2) The social predicament; and 3) The economic predicament.

The ecological predicament: why our ecosystems are breaking

Human dependence on nature

It may seem surprising (at least to an ecologist) that society continues to deny human dependence on nature. Yet this remains the case. Washington (2013, 2015) summaries some facts around this dependence, as listed below.

Food webs

Aldo Leopold (1949: 216) characterised ecosystems as a: 'fountain of energy flowing through a circuit of soils, plants and animals'. Energy is continually being added to ecosystems from the sun, being trapped by plants (producers) and these plants are eaten by herbivores, which may then be eaten by carnivores. The energy limits of the Earth's ecosystems cannot be ignored. Energy is life, but the amount coming to Earth is fixed. Humanity is now using about 12,000 times as much energy per day as was the case when farming first started

(Boyden 2004). So how much of the Earth's bioproductivity should be controlled by just one species? Vitousek et al. (1986) estimated that about 40% of net primary productivity (NPP) in terrestrial ecosystems was being co-opted by humans each year. Others have estimated figures somewhat higher or lower, but all are a huge percentage of the planet's NPP. The fact that 60% of ecosystem services are now being degraded or used unsustainably (MEA 2005) shows our current appropriation of NPP is too high. Clearly, we are way beyond what could be considered 'just' in terms of our fair share.

Keystone species

Some species have more effect on how energy moves through a food web, and even on what species are found in their communities. These are known as 'keystone species', important but little known parts of ecosystems. There are three types: 'predators', 'mutualists' and 'ecosystem engineers' (Cain et al. 2008). Keystone *predators* are often found at high levels in the food web, such as top predators like wolves, dingoes and jaguars. Keystone *mutualists* are organisms that participate in mutually beneficial interactions with other organisms, and their loss impacts strongly upon ecosystems. Keystone *ecosystem engineers* create habitat for other species, examples being grizzly bears (which move nutrients derived from eating salmon into the forest) and prairie dogs (whose tunnels create habitat for other species) (Washington 2013). It is thus critical we keep keystone species; the trouble is we don't know what many (perhaps most) of them are.

The nutrient cycles

Ecosystems require nutrients to be continually recycled. Essential nutrient cycles include those for phosphorus, sulphur, nitrogen and potassium. Ecosystem services are dependent on the balanced functioning of nature's nutrient cycles. However, humanity has pushed these cycles out of kilter, essentially doubling the amount of phosphorus and nitrogen moving through ecosystems, with major negative impacts (Washington 2013).

Overshoot

'Overshoot' is to shoot or pass over or beyond (Catton 1982). The term is relevant because it demonstrates that society has *exceeded ecological limits*, causing an environmental crisis. To understand overshoot (Washington 2018a), one needs to consider several factors, as covered below. Many environmental scientists, ecological economists and other scholars believe the overarching drivers of overshoot are the endless growth myth (and economy), overpopulation and overconsumption (as summarised in Washington 2015).

Ecosystem services

Gretchen Daily (1997) stated that ecosystem services are conditions and processes through which natural ecosystems: 'sustain and fulfil human life'. They maintain biodiversity and the production of ecosystem goods that include seafood, forage, timber, fibre and medicines. They embody the actual life-support functions that maintain society, such as cleansing and recycling. They also confer important aesthetic, spiritual and cultural benefits. Ecosystem services are commonly defined as the direct and indirect contributions of ecosystems to human well-being (De Groot et al. 2010). A total of 1,360 experts wrote the Millennium Ecosystem Assessment (MEA 2005). It split ecosystem services into four parts, being *provisioning services* (products obtained from ecosystems), *regulating services* (benefits obtained from regulation of ecosystem processes), *cultural services* (nonmaterial benefits) and *supporting services* (those necessary for the production of all other ecosystem services) (Figure 1.1).

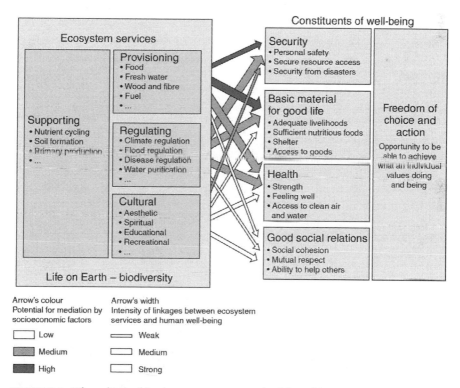

FIGURE 1.1 The relationships between ecosystem health and human well-being (MEA 2005)

The MEA (2005) noted that human use of all ecosystem services is growing rapidly. Overall it concluded that 60% of ecosystem services are being degraded or used unsustainably. Many ecosystem services are being degraded primarily to increase food supply. It should be emphasised, however, that ecosystem services are in fact anthropocentric, as they are defined as the services that nature provides to *humanity* (Washington 2019b). However, the rest of nature requires services from ecosystems as well. A narrow focus on ecosystem services can thus ignore that nature has intrinsic value, a right to exist for itself (Batavia and Nelson 2017).

Environmental indicators

Environmental indicators all show that society is in overshoot in regard to ecological limits. The *Ecological Footprint* (Wackernagel and Rees 1996) is based on the idea of 'biocapacity', the capacity of a given area to generate an ongoing supply of renewable resources and to absorb waste. The ecological footprint is monitored by the Global Footprint Network (GFN 2019). Ecological Footprint analysis compares human demands on nature with the biosphere's ability to regenerate resources and provide services. Unsustainability occurs if the area's Ecological Footprint exceeds its biocapacity. Both are usually expressed in global hectares (written as 'gha') or global hectares per person. In 2014, the Earth's total biocapacity was 12.2 billion gha, or 1.68 gha per person, while humanity's Ecological Footprint was 20.6 billion gha, or 2.84 gha per person (GFN 2019). Hence the Global Ecological Footprint is an unsustainable *1.7 Earths*. This is clearly not sustainable, as we must live on only one Earth. It should be noted however that this also assumes that no land is set aside for *other species* that consume the same biological material as humans (Stechbart and Wilson 2010). This is actually a dangerous assumption, as apart from any ethical 'right', nonhuman species provide the ecosystem services on which society depends (see Washington 2019b).

The Living Planet Index developed by the World Wide Fund for Nature (WWF) measures trends in thousands of vertebrate species populations, and shows a decline of 60% since 1970. In other words, the number of mammals, birds, reptiles, amphibians and fish across the globe is (on average) less than half what it was 40 years ago (WWF 2018). Biodiversity is declining in both temperate and tropical regions, but the decline is greater in the tropics. Habitat loss and degradation, and exploitation through hunting and fishing, are the primary causes of decline (Ibid.).

Planetary boundaries is a concept that seeks to define a 'safe operating space for humanity', and Rockstrom et al. (2009) identified that three planetary boundaries (out of nine) had already been exceeded – climate change, biodiversity extinction and nitrate pollution. Steffen et al. (2015) added phosphorus pollution. It is worth remembering however that 'planetary boundaries' as a term is also an anthropocentric concept. They are boundaries to define a 'safe operating space' *for humanity* and thus do not consider the rest of life on Earth. The concept however could be broadened to embrace this.

Extinction

Species extinction is 1,000 to 10,000 times above normal levels in the fossil record (Chivian and Bernstein 2008). Wilson (2003) believes the extinction rate is likely 10,000 times greater than normal, so that something like three species go extinct every hour. He concluded that humanity has become a serial killer of the biosphere. We are now in the midst of the sixth mass extinction we know of in the last 600 million years (Ceballos et al. 2015). Wilson (2003) has warned that without action, by the end of the century, *half* of all species on Earth may be extinct. More recently, a 2011 paper (Raven et al. 2011) suggested that (without action) *two-thirds* of terrestrial species were likely to become extinct by 2100. As Crist (2012: 150) concludes: 'We are losing our own family'. The latest UN report on extinction (IPBES 2019) notes that 1 million species are now threatened with extinction, many within decades. It notes that the average abundance of native species in most major land-based habitats has fallen by at least 20%, mostly since 1900. More than 40% of amphibian species, almost 33% of reef-forming corals and more than a third of all marine mammals are threatened. This report on extinction states the five direct drivers of extinction are: 1) changes in land and sea use; 2) direct exploitation of organisms; 3) climate change; 4) pollution and 5) invasive alien species (Ibid.). If we let this mass extinction happen, it would close off the evolutionary potential of more than half the living world. As Soulé and Wilcox (1980) observed: 'Death is one thing – an end to birth is something else'. The massive loss of living beings due to human action is what happens when society goes into ecological overshoot.

BOX 1.1 IN-THE-BLINK-OF-AN EYE WE SAVE HUMANITY FROM ITSELF, DOMINICK A. DELLASALA

This May, carbon emissions hit an all-time high as temperature gauges near the Arctic Circle in northwest Russia recorded the unthinkable: 87° F when it's supposed to be in the 50s! Perhaps, even worse, the Intergovernmental Science-Policy Platform on Biodiversity and Ecosystem Services released findings that 1 million species will soon go the way of the dodo, as humanity draws down more of its share of the planet's life-giving systems. This is a crisis of immense proportions and most of humanity really won't know what we've got till it's gone.

Every single species on this planet is a freak of the Universe. We were born out of a cosmic explosion billions of years ago triggering swirling stardust clouds and cosmic debris that eventually coalesced into this amazing and highly improbable (unique) blue ball of life.

With blatant disregard for the great Mystery in all this, we face an ultimatum.

Last year, the Alliance of World Scientists issued its second warning to humanity that we were on an unprecedented collision course with Nature. The Intergovernmental Panel on Climate Change issued its latest findings that we have just 12 years to drastically cut emissions or we will enter the never-been-seen-before world of climate chaos. So, for argument's sake, and to be on the safer side, let's start the countdown at 10 years before the climate is baked in. That's a blink-of-an-eye in a lifetime.

In crisis, there is always opportunity to heal, to rise up, to become a better person, to choose a different future, to be a better planet. We've all been there but it's time now for our species to rise up. To meet the challenge, I'm proposing a 'blink-of-an-eye' campaign to save humanity from itself. A moonshot-scale about face to keep dinosaur carbon in the ground by breaking our addiction to fossil fuels, and to keep atmospheric carbon in forests and other carbon-dense ecosystems by protecting them from insatiable developers.

And we need to combine ecological with social justice movements. The 'Green New Deal', The 'Global Deal for Nature', 'Nature Needs Half', 'Half Earth' have each set ambitious targets. '#MeToo', 'Black Lives Matter', Indigenous Rights, LGBTQ and other social change movements show the next generation that we at least gave it our best shot.

Such bold action on climate is being led by the youth – 'Our Children's Trust' and Greta Thunberg – along with elders like the International Council of Thirteen Grandmothers and the Man Kind Project seeking to heal interpersonal and planetary wounds.

Carl Sagan prophetically wrote in *Contact*, 'You're an interesting species. An interesting mix. You're *capable of such* beautiful dreams, and such horrible nightmares'. As a scientist and a loving parent, those words ring truer to me today than in 1970. I'm doing my part to make sure the beautiful dreams prevail over the horrible nightmares because it can all be gone in the-blink-of-an-eye.

For more information:
http://scientistswarning.forestry.oregonstate.edu/; www.ipbes.net/news/ Media-Release-Global-Assessment; www.forestlegacies.org/; www .congress.gov/116/bills/hres109/BILLS-116hres109ih.pdf; www.leonardodica prio.org/the-global-deal-for-nature/; www.half-earthproject.org/.

Ecosystem collapse

Ultimately, human existence depends on maintaining the rich web of life within which we evolved (Washington 2013). The rapidity of current environmental change is unique in human history (MEA 2005; Gowdy et al. 2010; Wijkman and Rockstrom 2012). Many entire ecosystems are degraded and verging on collapse. This situation has been described as the 'rivet popper' analogy by Ehrlich and Ehrlich (1981). If you keep taking out rivets from an aeroplane, at some point it will fall apart and crash. If you keep removing species, so will ecosystems. Ecosystem collapse is discussed in more detail in Chapter 10.

Being way past ecological limits

We need to understand that a key part of our predicament is that society is *way past* ecological limits. I will not discuss overpopulation in detail here as this is covered in Chapter 4. During the 20th century (data mostly from Rees 2008):

- Human population went up 4-fold to 6.4 billion.
- Industrial pollution went up 40-fold.
- Energy use increased 16-fold and CO_2 emissions 17-fold.
- Fish catches expanded 35-fold.
- Water use increased 9-fold.
- Mining of ores and minerals grew 27-fold (UNEP 2011).
- One quarter of coral reefs were destroyed and another 20% degraded (MEA 2005).
- 35% of mangroves were lost (in just two decades) (MEA 2005).
- At least half of all wetlands were lost (Meadows et al. 2004).

The above illustrates why all environmental indicators are in decline. We should note that humans use 75% of the Earth's land mass (excluding Greenland and Antarctica) (Erb et al. 2009). Some 18.7 million hectares of forest are cleared each year worldwide (WWF n.d.). Half the world's people live in countries where water tables are falling as aquifers are depleted (Brown 2011). Regarding climate change, it has been said that most species extinction models for climate change indicate alarming consequences for biodiversity, with worst-case scenarios leading to extinction rates that would qualify as the sixth mass extinction in the history of the Earth (Bellard et al. 2012). IPBES (2019) lists climate change as the third key cause of extinction. There are of course other environmental impacts of climate change on top of extinction, such as rising sea level, increasing extreme weather, major impacts on water availability and food production, and ocean acidification changing marine ecosystems (Pittock 2009; IPCC 2014). I will not delve extensively into the climate crisis, given that this has been much discussed (e.g. Pittock 2009; IPCC 2014, 2018). The IPCC (2014, 2018) has developed many graphics to explain the impacts of climate change, but these are quite complex. The easiest simple graphic to aid understanding of the impacts of climate change still seems to be that produced by Stern (2006), and shown below in Figure 1.2. Climate change will impact strongly on food production, water systems, native ecosystems and increase extreme weather. As climate change accelerates, the risk of abrupt and likely irreversible changes increases. The reality is that human-caused climate change is changing the world rapidly – to the great disadvantage of the ecosystems society relies on, and hence to our own great

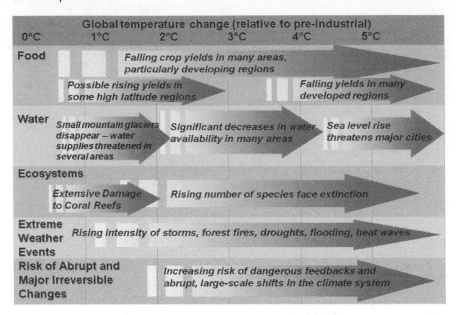

FIGURE 1.2   Projected impacts of climate change. (Source: Stern 2007 'The Economics of Climate Change', talk given to the World Bank, 23 March 2007. Figure based on figure 2 of 2006)

personal disadvantage (and that of our children). If society pushes climate into more than two degrees of warming (as seems likely, Spratt and Dunlop 2018), then we could enter runaway climate change, causing even greater mass extinction, a food crisis and even the collapse of society as we know it. Indeed, we run the risk of turning our abundant Earth into an uninhabitable Earth (Wallace-Wells 2019).

All the above discussion shows us that we are *bankrupting* nature (Wijkman and Rockstrom 2012). On a finite world with expanding population and consumption, clearly something has got to give. In the 'World Scientists Warning to Humanity: A second notice' Ripple et al. (2017) warn:

> By failing to adequately limit population growth, reassess the role of an economy rooted in growth, reduce greenhouse gases, incentivize renewable energy, protect habitat, restore ecosystems, curb pollution, halt defaunation, and constrain invasive alien species, humanity is not taking the urgent steps needed to safeguard our imperilled biosphere.

As Wijkman and Rockstrom (2012) conclude, we are consuming the past, present and future of our biosphere. So society *does* indeed have a major problem, which is why we desperately need to heal the world.

The social predicament: why our society is broken

What is fair? Equity and equality

A key part of living sustainably into the future is about 'fairness'. The words 'equity' and 'equality' are much discussed in the literature. 'Equity' is the quality of being fair and impartial. 'Equality' is the state of being equal, especially in status, rights, income and opportunities (Washington 2015). Widening income gaps and accelerated ecological change suggest that the mainstream global community still pays little more than lip-service to the sustainability ideal (Moore and Rees 2013). Another way to think about it is in terms of 'sharing'. How should we live? Should just a few have most of the land, food and wealth? Or should we share things so that everyone has a decent quality of life (Washington 2015)?

Almost all tribal cultures had a system of redistribution of wealth, from Chinook 'potlatch' to Mosaic 'year of jubilee' (Czech 2013). The point is that both conservation of resources and redistribution of wealth are essential for sustainability (Ibid.). Ethically, there is no debate, and most people profess to agreeing with the idea of 'sharing', though many also slip into denial. Wallich (1972) of the US Federal Reserve has argued: 'growth is a substitute for equality of income. So long as there is growth there is hope, and that makes large income differential tolerable'. It suggests all we have to do is bake a bigger cake. However, we *can't* in fact bake a bigger cake when the 'oven' (the ecosystems of the world) is a fixed size, and actually degrading, as discussed earlier. Ophuls (1992) observes that growth is a false substitute for equity. We need to be fairer in how we distribute the cake we have now. Yet currently we fail to do so, and this is getting worse.

BOX 1.2 SUSTAINABILITY NEEDS EQUALITY, RICHARD G. WILKINSON AND KATE PICKETT

The climate breakdown caused by the current one degree C of global warming is already causing hundreds of thousands of deaths each year. With world carbon emissions continuing to rise, climate scientists fear we are heading for four or perhaps five degrees of warming by the time the present cohort of infants reach old age. And no one knows how far we are from much more widespread disaster, perhaps including runaway global warming, and social and economic collapse.

To avoid inflicting appalling difficulties on the later lives of the children with us now, there are three components of a solution that we must pursue. First, we do of course need massive technical change – to renewable sources of power, to minimise the use of plastics, to abandon petrol and diesel vehicles, to making goods which can be repaired and recycled, etc. Second, we need taxes and subsidies to alter market choices and encourage people to eat much less meat, to use public transport, to replace flying with conferencing on-line, and so on. But third, the transformation to sustainability must be supported by institutional change.

We have been led into this crisis by structures which make almost everyone want to maximise their income and consumption and every business want to maximise their sales and profits. There could hardly be a worse place from which to start the transition to sustainability. The desire to maximise income and consumption leads almost inevitably to widespread popular opposition to environmental taxes. The Gilets Jaunes protests in France were an example. They started as a protest against a planned tax increase on motor vehicle fuel intended to reduce carbon emissions, but the protests were so strong and so extensive that the government of President Macron bowed to public pressure and withdrew them. For their part, corporations wanting to maximise profit will continue to attack climate science wherever it threatens their interests and will endlessly try to avoid making the changes which sustainability requires.

Unless we overcome these extraordinarily powerful and damaging forces we will fail to save the planet and ourselves. But there are powerful policy levers at our disposal. We have shown in our two most recent books (*The Spirit Level* (2009) and *The Inner Level* (2019)) that reducing income differences between rich and poor substantially reduces consumerism. By making class and status less important it reduces the importance of money as the means of expressing self-worth. People in more equal societies are less likely to spend money on status goods. Less out for themselves, they are less likely to get into debt or to work longer hours. Community life strengthens and, as a result, people become more willing to act for the common good.

One of the best ways of reducing income differences and embedding greater equality into society is to increase economic democracy. We should legislate for substantial employee representation on company boards and incentivise the development of employee-owned businesses and employee cooperatives. More democratic business models have many advantages: they reduce employee income differences, increase productivity and are more public spirited.

For more information:
www.equalitytrust.org.uk/
Wilkinson, R. and Pickett, K. (2010) *The Spirit Level: Why Equality is Better for Everyone*, London: Penguin.
Wilkinson, R. and Pickett, K. (2019) *The Inner Level: How More Equal Societies Reduce Stress, Restore Sanity and Improve Everyone's Well-being*, London: Penguin.

The modern world has not brought greater fairness as promised; quite the opposite. Only 16% of people live in the developed world, yet account for 78% of global consumption expenditure. Meanwhile, 40% of the world's population struggles to subsist on less than $2 a day (Dietz and O'Neill 2013). As a global society, we are becoming *less fair*, not more. The average member of the top 1%

of the world's population is almost 2,000 times richer than someone from the poorer half (Renner 2012). Rather than ending poverty, modernism has led to the environmental crisis, and this is now impacting hugely on the poor, who rely on free ecosystem services to survive (Washington 2015). Inequity in income distribution has increased within OECD counties but also in major emerging economies such as China and India (OECD 2011). Growth is continuing, but the poor get less and less of the benefits (Layard 2005). For every $100 of global economic growth between 1990 and 2001, only 60 cents went to people with incomes of less than $1 a day (Dietz and O'Neill 2013). Governments did not actually 'plan' to lower social cohesion or to increase violence, physical and mental ill-health, or drug abuse. However, inequality of income increases all of these (Wilkinson and Pickett 2010). It is time to tackle the core problem of inequality, for community and equality are mutually reinforcing (Ibid.).

The mantra of development, growth and 'trickle down' (that we have been told to believe in) is thus not actually helping the poor. Someone is profiting from economic growth, but it is not the poor. Kopnina and Blewitt (2015) note that the 'trickle-down effect' is a myth, as much of the 'wealth' that was created in the good times was not real – just numbers on a screen – and it is the rich rather than the poor who have benefited. Sachs (2002) explains that in the neo-liberal ideological view, greater 'equity' (if it matters at all) is seen as expanding the reign of the economic law of 'demand and supply' across the globe. Hence the endless growth economy has exacerbated inequality (Washington 2015).

In their illuminating book *The Spirit Level*, Wilkinson and Pickett (2010) explain that 'inequality is bad for everyone'. They summarise extensive research showing that inequality of income worsens a whole range of social aspects. These include health, mental illness, violence, homicides, obesity, drug use, life expectancy, competitive consumption, trust, children's educa tional performance, teenage births, imprisonment rates and social mobility. They develop an index of health and social problems, shown in Figure 1.3. All these social indicators are worse in societies with higher income inequality. This is because humanity is an intensely social species, so what matters is where we stand in relation to others in our own society. Human beings are driven to preserve the 'social self' and are vigilant to threats that may jeopardise social esteem or status. Increased inequality ups the stakes in the competition for status, so then status matters even more. For a species that thrives on friendship, and enjoys cooperation and trust (and which has a strong sense of fairness), it is clear that social structures based on inequality, inferiority and social exclusion must inflict a great deal of social pain (Wilkinson and Pickett 2010). Inequality erodes empathy, hardens hearts, undermines public trust, incites violence, saps our imagination and destroys the public spirit that upholds democracy (Orr 2013). As a result, modern society can be said to be *broken* and likely (without change) to become increasingly dysfunctional.

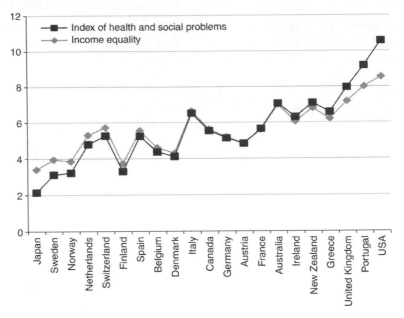

FIGURE 1.3 Index of health and social problems vs. income inequality in world nations. Drawn based on data from Wilkinson, R. and Pickett, K. (2010) *The Spirit Level: Why Equality is Better for Everyone.*

Consumerism is discussed in Chapter 5, and this acts to increase inequality. A final point to consider is about how far one extends the idea of 'fairness'. Is this limited just to 'our social group', does it apply to the whole of society, or should we extend the idea of fairness and justice to future generations *and* to all of nature as well? Certainly, the idea of justice has for centuries been seen as being limited to human society, and only recently have some argued for 'ecojustice', being justice for nature (see Chapter 2).

Social cohesion and social capital: necessary so we can act

If we are to achieve social sustainability, then society has to function cohesively to solve problems. We need social *cohesion*, social *inclusion* and social *capital* (Washington 2015). Social cohesion is a somewhat vague concept, but has been said to describe social connectedness, including family and community well-being (Hulse 2005). It is developed through the nature and quality of interrelationships between people. It can be said to manifest in the 'sense of community'. Societies may lose cohesion if people don't feel 'included'. Social inclusion is the provision of certain rights to all individuals and groups, such as employment, adequate housing, health care, education and training. If they feel included then they feel part of a community, and social cohesion is greater. Such ideas are not new, for Adam Smith (1759) believed an economy required a 'social bond' that brought social stability.

Social capital is the expected benefits derived from the cooperation between individuals and groups. The core idea is that social networks have value. Robert Putnam (2000) stated it refers to the collective value of all 'social networks' and the inclinations that arise from these networks to do things for each other. Fukuyama (2002) defines it as shared norms or values that promote social cooperation. Social capital is seen as a key component to maintaining democracy. It is a producer of 'civic engagement', a broad societal measure of communal health (Alessandrini 2002). Social capital can be measured by the amount of trust and 'reciprocity' in a community. Civic engagement is a manifestation of social capital. However, is social capital increasing or decreasing? Putnam (2000) believes social capital is declining in the United States, demonstrated by less trust in government and less civic participation. Television and urban sprawl play a significant role in making the United States less 'connected' (Ibid.). Others however use different assessments and question whether this is the case (Clark 2015). Like so many topics in academia, it is a contested topic. However, any decline in social capital is unlikely to help us heal the world.

The economic predicament: why our economy is broken

The causes of our broken economy are discussed further in Chapter 3 in regard to the unsustainability of 'growthism', and in Chapter 5 in regard to the unsustainable assumptions of neoclassical economics. At its simplest, however, the economy is how we organise things in our society, how we produce food and materials, distribute and trade them, and swap skills. Economics is thus the study of how humans make their living, how they satisfy their needs and desires (Common and Stagl 2005). The economy was thus always meant to *serve* society. This is no longer the case today. 'Good' economics should be good 'management of the home' (its meaning in ancient Greek). However, this is now anything but the case. Modern economics is fraught with issues of worldview, ideology, assumptions, ignorance of ecological reality and denial (Daly 1991, 2014; Rees 2019). Denying the problems of the 'growth economy' is also arguably the largest 'elephant in the room' that most in society refuse to see. It should be emphasised that though we talk about the economy as if it is real, it is just something society has constructed as an idea (Rees 2019). You cannot see the economy from space or even from a plane. We can of course see the negative effects (e.g. deforestation, desertification) of an addiction to an endless growth economy on a finite planet (see Chapter 3 re 'growthism'). We should also be aware that every economy is in fact an ethical economy – to the extent that the way we structure our economy is defined by our worldview and ethics (see Chapter 2).

Kopnina and Blewitt (2015) note that sustainability cannot be met by simply tinkering with the current economic system. To talk about 'economic sustainability' one must confront the issue that our current economy is fundamentally broken and in need of renewal (Jackson 2009). The growth economy has become the 'given truth' of our times, the 'chief sacred' in society (Ellul 1975). More than anything else in our society, the growth economy remains taboo to question, and the silence of denial concerning it reigns almost supreme. This is true even in many academic, media and environmental circles. It has been argued by Sukhdev (2013: 143) that there is an emerging consensus among governments and business leaders that: 'all is not well with the market–centric economic model' that dominates today. Similarly, Alperovitz (2014) observes that the current political and economic model is incapable of addressing the big sustainability challenges we face. The key problem with neoclassical economics is that its fundamental premise of 'endless growth' is both irrational and unsustainable when one lives on a finite planet (that is already past ecological limits). Hence our current economics is not just broken – it is fundamentally unsustainable, degrading the natural world that supports society. Accordingly, it is driving us towards both environmental and social collapse. Sadly, we have constructed our economy optimally badly, in that it is based on consuming and bankrupting nature (Wijkman and Rockstrom 2012). Changing from an endless growth economy to a steady state economy (Daly 1991, 2014) is thus a key part of healing our world. Growth economics and its assumptions are further discussed in Chapters 3 and 5.

BOX 1.3 THE STEADY STATE ECONOMY: THE SUSTAINABLE ALTERNATIVE TO GROWTH, BRIAN CZECH

Economic growth is an increase in the production and consumption of goods and services in the aggregate. It entails increasing population times per capita consumption, higher throughput of materials and energy, and a growing ecological footprint. Economic growth is indicated by increasing gross domestic product (GDP)

Economic growth is distinguished from 'economic development', which refers to qualitative change independent of quantitative growth. For example, economic development may refer to the attainment of a more equitable distribution of wealth, or a sectoral readjustment reflecting the evolution of consumer preference or newer technology.

The size of an economy may undergo one of two trends: growth or recession. Otherwise it is stable, in which case it is a 'steady state economy'. Theoretically and temporarily, a steady state economy may have a growing population with declining per capita consumption, or vice versa, but neither of these scenarios is sustainable in the long run. Therefore, 'steady state economy' connotes constant populations of people (and, therefore, 'stocks' of labor) and constant stocks of capital. It also has a constant rate of throughput, i.e. energy and materials used to produce goods and services.

Within a given technological framework these constant stocks will yield constant flows of goods and services. Technological progress may yield a more efficient 'digestion' of throughput, resulting in the production of more (or more highly valued) goods and services. However, there are limits to productive efficiency imposed by the laws of thermodynamics and therefore limits to the amount and value of goods and services that may be produced in a given ecosystem. In other words, there is a maximum size at which a steady state economy may exist. Conflicts with ecological integrity and environmental protection occur long before a steady state economy is maximised.

'Constancy' of population and capital stocks does not imply absolutely unchanging population and capital stocks at the finest level of measurement. Rather, 'constant' implies mildly fluctuating in the short run but exhibiting a stable equilibrium in the long run. Long-run changes reflect evolutionary, geological or astronomical processes that alter the carrying capacity of the earth for the human economy. Dramatic examples include atmosphere-altering volcanoes and massive meteorite collisions.

Just as economic growth is the predominant macroeconomic policy goal identified or implied by neoclassical economics, the steady state economy is the predominant macroeconomic policy goal identified or implied by ecological economics. To the extent that ecological economics is a normative transdisciplinary endeavor rather than a purely analytical framework, its three main concerns are sustainability, equity, and efficiency, each of which may be served via public policy. Neither economic growth nor economic recession are sustainable; therefore, the steady state economy remains the only sustainable prospect and the appropriate policy goal for the sake of sustainability.

For more information:

Support for a steady state economy is gauged at the Center for the Advancement of the Steady State Economy, www.steadystate.org, which currently has 14,000 signatories. You too may choose to sign the position statement?

There is discussion of neoliberalism in this book in terms of our economy. As this term can mean different things to different people I should clarify that I use the definition given by *The Handbook of Neoliberalism* (Springer et al. 2016: 2):

> At a base level we can say that when we make reference to 'neoliberalism', we are generally referring to the new political, economic and social arrangements within society that emphasize market relations, re-tasking the role of the state, and individual responsibility. Most scholars tend to agree that neoliberalism is broadly defined as the extension of competitive markets into all areas of life, including the economy, politics and society.

My concern regarding neoliberalism is precisely because it privileges the market above ethics, above sustainability and, indeed, (as we shall see) above survival. Both the economy and the market are ideas, neither thinks about what is wrong with the world, neither cares for society nor nature. A dominant neoliberal ideology in society is thus not aiding the healing of the environmental crisis.

Conclusion

The detail above shows there are good grounds for arguing that both our society and economy are broken and unsustainable. There are equally good grounds for arguing that the nature that supports society is currently breaking. We are in a major mass extinction event, where we may send half of life on Earth (or more) extinct. The predicament of society is thus very real – if one is willing to look at the facts and not retreat into denial, which is a distressingly common occurrence (see Chapter 3). A sustainable society has to be sustainable ecologically, socially and economically (Washington 2015), yet we are failing in regard to all three. This is not a reason for despair or denial. Rather, it should be a spur to *heal all three*.

What can *I* do?

Society's predicament is very real. What can each of us do about it? There are many solutions we can embrace in regard to the themes of ecological, social and economic sustainability.

In regard to healing to create *ecological sustainability*, there are many things you can do:

1) Acknowledge, understand and inform others about society's predicament. Society faces difficult problems, but the first step to solving them is *accepting reality*, and then acting to change things.
2) Understand and explain that we depend on nature to survive; there is no 'planet B' to move to (Brown 2011).
3) Do not take nature for granted. Ignore it and it will indeed go away. Indeed environmental indicators show that this is underway.
4) Adopt an ecocentric worldview and ethics (see Chapter 2) that acknowledge that nature has intrinsic value, deserves respect and that humanity has a duty of care towards nature (Washington et al. 2017; Washington 2018b).
5) Act to solve the climate crisis by helping to decarbonise our society within two decades. This means moving immediately to renewable energy and energy efficiency (see Chapter 6).
6) Conserve nature and support the 'Half Earth' vision (Wilson 2016) where half of terrestrial lands are protected. This means no further clearing of natural ecosystems; it means extensive ecological restoration; it means greater connectivity between natural areas; it means curtailing huge developments and controlling urban sprawl. Essentially, it also means controlling population

growth (see Chapter 4) and overconsumption (see Chapter 5). Healing the planet needs to be our core vision, not continuous unsustainable 'development'.

7) Lobby to keep our large natural areas (wilderness), national parks and other reserves sacrosanct, as these are the keystone foundations of nature conservation (Kopnina and Washington 2019).

8) Lobby to stop further toxification of the planet. This means both limiting resource throughput and strictly controlling the 100,000 chemicals in use (Rockstrom et al. 2009) and the creation of the 1,000 or so new synthetic chemicals added each year (ATDSR 2010).

9) Support ecologically sustainable (or 'regenerative') farming. This means a return to organic farming that minimises artificial fertiliser and biocide use, and uses Integrated Pest Management (EPA 2014). If done properly this may be able to produce as much food as conventional agriculture (Wijkman and Rockstrom 2012). It means redesigning our agriculture to be truly ecologically sustainable (Hill 2014). If possible, buy locally grown and organic food, and avoid foods/beverages transported for long distances.

10) Argue that all the plants and animals we use (and dismiss as 'resources') must be harvested in an ecologically sustainable way. This means not overcutting our forests or overharvesting our fisheries (as we currently do). Don't buy overseas rainforest timber, don't use palm oil (or buy products containing it) whose production generally means destruction of rainforests.

11) Eat less meat. This doesn't have to mean becoming a vegan, it can mean eating less of the meat that causes greater environmental impact. Impact declines from lamb and beef to pork to chicken (EWG 2011; Wilson 2015). Rangeland beef and lamb has less impact than that produced in a feedlot.

In regard to *social sustainability*, you can:

12) Argue that the highest social values are sustainability, efficiency, sufficiency, equity and equality, beauty and community (Meadows et al. 2004).

13) Argue for a vision farsighted enough, flexible enough and wise enough not to undermine either its natural or its social systems of support (Meadows et al. 2004).

14) Argue for improving society's *quality of life* (rather than a 'standard of living' based on the level of wealth, comfort and material goods) through factors such as cultural expression, political freedoms and civil rights, and not just increasing the GDP (Heinberg 2011).

15) Speak out for greater commitment to social cohesion, social inclusion and social capital.

16) Argue for greater equity between rich and poor nations. Rich nations should give greater overseas aid to poorer nations for ecologically sustainable projects that help heal the world (not promote growth). Argue also for greater equity and equality of income within nations.

17) Support experiments with every form of *economic democracy* – employee ownership, producer and consumer cooperatives, employee representatives on boards, etc. (Wilkinson and Pickett 2010). This means a greater focus on 'cooperation' rather than competition. Successful cooperatives such as Mondragon in Spain emphasise cooperation, as opposed to most corporations that focus on competition (Dietz and O'Neill 2013).

18) Engage more in environmental politics in your democracy. Democracy needs to be widened beyond the major party system. We need a move to more *participatory democracy*, where the general public is more directly involved in decision-making (Elster 1998). However, to broaden it further, we also need to move beyond that to *ecodemocracy*, where nature is given voting proxies in our decision-making systems (see www.ecodemocracy.net).

19) Speak out in support of a shift in public values, so that instead of inspiring 'admiration and envy', conspicuous consumption is seen as a sign of 'greed and unfairness' that damages both society and the planet (Wilkinson and Pickett 2010: 270).

In regard to an *ecologically sustainable economy*, there are many things you can do (see also Washington 2017). Here are two key points (but see Chapter 5 for detailed actions):

20) Challenge the mantra of the growth economy and consumerism. Point out there is another way – a *steady state economy* based on: a) an ecologically sustainable population; b) much less use of resources; c) greater equity (Daly 2014). Sign the CASSE position statement at https://steadystate.org/act/sign-the-position/read-the-position-statement/ and if one exists near you, join a local chapter. Discover the many positive steps we can take towards a steady state economy (Washington 2017).

21) Speak out for tax-shifting, by taxing the 'bads' that degrade our environment. Speak out also for subsidy-shifting, especially taking the subsidies given to fossil fuels and transferring these to renewable energy. There are at least $700 billion worldwide of bad subsidies given to damaging activities (Brown 2011). Indeed, governments globally spend $640 billion just on fossil fuel subsidies (OECD 2015).

There are thus many things we each can do to heal the world and move towards ecological, social and economic sustainability. Of course, you may not be able to do all these, but you *can* certainly act on some of them and thereby help to heal the world.

References

Alessandrini, M. J. (2002) 'Is civil society an adequate theory?', *Third Sector Review*, **8** (2): 105–119.

Alperovitz, G. (2014) 'The political-economic foundations of a sustainable system', in *State of the World 2014: Governing for Sustainability*, ed. L. Mastny, Washington: Island Press, pp. 191–202.

ATDSR (2010) '"Chemicals, cancer and you" booklet, agency for toxic substances and disease registry', see: www.atsdr.cdc.gov/emes/public/docs/Chemicals,%20Cancer,%20and%20You%20FS.pdf (accessed 15/6/14)

Batavia, C. and Nelson, M. P. (2017) 'For goodness sake! What is intrinsic value and why should we care?', *Biological Conservation*, **209**: 366–376.

Bellard, C., Bertelsmeier, L. P. et al (2012) 'Impacts of climate change on the future of biodiversity', *Ecology letters*, **15** (4): 365–377.

Boyden, S. (2004) *The Biology of Civilisation: Understanding Human Culture as a Force in Nature*, Sydney: UNSW Press.

Brown, L. (2011) *World on the Edge: How to Prevent Environmental and Economic Collapse*, New York: W.W. Norton and Co.

Cain, M., Bowman, W. and Hacker, S. (2008) *Ecology*, Massachusetts: Sinauer Associates.

Cardinale, B. J., Duffy, E., Gonzalez, A. et al (2012) 'Biodiversity loss and its impact on humanity', *Nature*, **486** (7401): 59–67, DOI: http://dx.doi.org/doi:10.1038/nature11148.

Catton, W. (1982) *Overshoot: The Ecological Basis of Revolutionary Change*, Chicago: University of Illinois Press.

Ceballos, G., Ehrlich, A. and Ehrlich, P. (2015) *The Annihilation of Nature: Human Extinction of Birds and Mammals*, Baltimore: Johns Hopkins University Press.

Chivian, E. and Bernstein, A. (2008, eds) *Sustaining Life: How Human Health Depends on Biodiversity*, Center for Health and the Global Environment, New York: Oxford University Press.

Clark, A. (2015) 'Rethinking the decline in social capital', *American Politics Research*, **43** (4): 569–601.

Common, M. and Stagl, S. (2005) *Ecological Economics: An Introduction*, Cambridge: Cambridge University Press.

Crist, E. (2012) 'Abundant Earth and the population question', in *Life on the Brink: Environmentalists Confront Overpopulation*, eds. P. Cafaro and E. Crist, Georgia: University of Georgia Press, pp. 141–151.

Czech, B. (2013) *Supply Shock: Economic Growth at the Crossroads and the Steady State Solution*, Canada: New Society Publishers.

Daily, G. (1997) *Natures Services: Societal Dependence on Natural Ecosystems*, Washington: Island Press.

Daly, H. (1991) *Steady State Economics*, Washington: Island Press.

Daly, H. (2014) *From Uneconomic Growth to the Steady State Economy*, Cheltenham: Edward Elgar.

De Groot, R., Fisher, B. and Christie, M. (2010) 'Integrating the ecological and economic dimensions in biodiversity and ecosystem service valuation', in *The Economics of Ecosystems and Biodiversity: Ecological and Economic Foundations*, ed. P. Kumar, London: Earthscan, pp. 9–40.

Dietz, R. and O'Neill, D. (2013) *Enough is Enough: Building a Sustainable Economy is a World of Finite Resources*, San Francisco: Berrett-Koehler Publishers.

Ehrlich, P. and Ehrlich, A. (1981) *Extinction: The Causes and Consequences of the Disappearance of Species*, New York: Random House.

Ellul, J. (1975) *The New Demons*, New York: Seabury Press.

Elster, J. (1998, ed) *Deliberative Democracy (Cambridge Studies in the Theory of Democracy)*, Cambridge, UK: Cambridge University Press.

EPA. (2014) 'Integrated Pest Management (IPM) Principles', see: www.epa.gov/safepest control/integrated-pest-management-ipm-principles (accessed 15 March 2019)

Erb, K., Haberl, H., Krausmann, F. et al (2009) 'Eating the planet: Feeding and fueling the world sustainably, fairly and humanely – A scoping study', Social ecology working paper no. 116, Vienna: Institute of Social Ecology and Potsdam Institute for Climate Impact Research. see: www.researchgate.net/publication/260403612_Eating_the_Planet_Feedin g_and_fuelling_the_world_sustainably_fairly_and_humanely_-_a_scoping_study (accessed 27 March 2019).

EWG. (2011) 'Meat eater's guide to climate change and health', Environmental Working Group. see: http://static.ewg.org/reports/2011/meateaters/pdf/methodology_ewg_ meat_eaters_guide_to_health_and_climate_2011.pdf (accessed 27 March 2019).

Fukuyama, F. (2002) 'Social capital and development: The coming Agenda', *SAIS Review*, **22** (1): 23–37.

GFN. (2019) 'Country trends', Global Footprint Network. see: http://data.footprintnet work.org/#/countryTrends?cn=5001&type=BCpc,EFCpc.

Gowdy, J., Howarth, R. and Tisdell, C. (2010) 'Discounting, ethics and options for maintaining biodiversity and ecosystem integrity', in *The Economics of Ecosystems and Biodiversity: Ecological and Economic Foundations*, ed. P. Kumar, London: Earthscan, pp. 257–284.

Heinberg, R. (2011) *The End of Growth: Adapting to Our New Economic Reality*, Canada: New Society Publishers.

Hill, S. B. (2014) 'Considerations for enabling the ecological redesign of organic and conventional agriculture: A social ecology and psychosocial perspective', in *Organic Farming, Prototype for Sustainable Agricultures*, eds. S. Bellon and S. Penvern, Dordrecht: Springer, pp. 401–422.

Hulse, K. (2005) *Housing, Housing Assistance and Social Cohesion in Australia*, Melbourne: Monash Research Centre Final Report.

IPBES. (2019) 'Nature's dangerous decline "unprecedented" species extinction rates "accelerating"', Press release Intergovernmental Science-Policy Platform on Biodiversity and Ecosystem Services. see: www.ipbes.net/news/Media-Release-Global-Assessment.

IPCC. (2014) 'Summary for policymakers, fifth assessment report, international panel on climate change', see: http://ipcc-wg2.gov/AR5/images/uploads/WG2AR5_SPM_FINAL. pdf (accessed 16 March 2018).

IPCC. (2018) *Global Warming of 1.5˚ C*. International Panel on Climate Change. see: www. ipcc.ch/pdf/special-reports/sr15/sr15_spm_final.pdf (accessed 16 March 2019).

Jackson, T. (2009) *Prosperity without Growth: Economics for a Finite Planet*, London: Earthscan.

Kopnina, H. and Blewitt, J. (2015) *Sustainable Business: Key Issues*, London: Routledge.

Kopnina, H. and Washington, H. (2019) *Conservation: Integrating Social and Ecological Justice*, New York: Springer.

Layard, R. (2005) *Happiness: Lessons from a New Science*, New York: Penguin Press.

Leopold, A. (1949) *A Sand County Almanac*, New York: Ballantine Books.

MEA. (2005) *Living Beyond Our Means: Natural Assets and Human Wellbeing, Statement from the Board*. Millennium Ecosystem Assessment, United Nations Environment Programme (UNEP). see: www.millenniumassessment.org/documents/document.429.aspx.pdf (accessed 19 February 2018).

Meadows, D., Randers, J. and Meadows, D. (2004) *Limits to Growth: The 30-year Update*, Vermont: Chelsea Green.

Moore, J. and Rees, W. (2013) 'Getting to one-planet living', in *State of the World 2013: Is Sustainability Still Possible?*, ed. L. Starke, Washington: Island Press, pp. 39–50.

OECD. (2011) 'Towards green growth', Organisation for Economic Co-operation and Development. see: www.oecd.org/env/towards-green-growth-9789264111318-en.htm (accessed 10 March 2019).

OECD. (2015) 'Towards green growth: Tracking progress', Organisation for Economic Co-operation and Development. see: https://read.oecd-ilibrary.org/environment/towards-green-growth_9789264234437-en#page1 (accessed 27 March 2019).

Ophuls, W. (1992) *Ecology and the Politics of Scarcity Revisited: The Unraveling of the American Dream*, New York: W. H. Freeman.

Orr, D. (2013) 'Governance in the long emergency', in *State of the World 2013: Is Sustainability Still Possible?*, ed. L. Starke, Washington: Island Press, pp. 279–291.

Pittock, A. B. (2009) *Climate Change: the Science, Impacts and Solutions*, Australia: CSIRO Publishing/Earthscan.

Putnam, R. (2000) *Bowling Alone: the Collapse and Revival of American Community*, New York: Simon and Schuster.

Raven, P., Chase, J. and Pires, J. (2011) 'Introduction to special issue on biodiversity', *American Journal of Botany*, **98**: 333–335.

Rees, W. (2008) 'Toward sustainability with justice: Are human nature and history on side?', in *Sustaining Life on Earth: Environmental and Human Health through Global Governance*, ed. C. Soskolne, New York: Lexington Books, pp. 81–94.

Rees, W. (2019) 'End game: the economy as eco-catastrophe and what needs to change', *Real World Economic Review*, **87**: 132–148.

Renner, M. (2012) 'Making the green economy work for everybody', in *Moving toward Sustainable Prosperity: State of the World 2012*, ed. L. Starke, Washington: Island Press., pp. 3–21

Ripple, H., Wolf, C., Newsome, T. et al (2017) 'World scientists' warning to humanity: A second notice', *Bioscience*, **67** (12): 1026–1028. see: https://science.gu.se/digitalAssets/1671/1671867_world-scientists-warning-to-humanity_-a-second-notice_english.pdf (accessed 27 March 2019).

Rockstrom, J. et al (2009) 'Planetary boundaries: Exploring the safe operating space for humanity', *Ecology and Society*, vol. 14, no. 2, p. 32. see: www.ecologyandsociety.org/vol14/iss2/art32/ (accessed 21 March 2019).

Sachs, W. (2002) 'Ecology, justice and the end of development', in *Environmental Justice, Discourses in International Political Economy – Energy and Environmental Policy*, ed. J. Byrne, vol. 8, New Brunswick, USA & London, UK: Transaction Publishers, pp. 19–36.

Smith, A. (1759) *The Theory of Moral Sentiments*, London: A. Millar.

Soulé, M. E. and Wilcox, B. A. (1980) 'Conservation biology: Its scope and its challenge', in *Conservation Biology: An Evolutionary-Ecological Perspective*, eds. M. Soulé and B. Wilcox, Sunderland (MA): Sinauer, pp. 1–8.

Spratt, D. and Dunlop, I. (2018) *What Lies Beneath: The Scientific Understatement of Climate Risks*, Melbourne, Australia: Breakthrough - National Centre for Climate Restoration. See: https://climateextremes.org.au/wp-content/uploads/2018/08/What-Lies-Beneath-V3-LR-Blank5b15d.pdf.

Springer, S., Birch, K. and MacLeavy, J. (2016) *The Handbook of Neoliberalism*, First edition, London: Routledge.

Stechbart, M. and Wilson, J. 2010. *Province of Ontario Ecological Footprint and Biocapacity Analysis*, Oakland, CA: Global Footprint Network. See: www.footprintnetwork.org/con

tent/images/uploads/Ontario_Ecological_Footprint_and_biocapacity_TECHNICAL_report.pdf (accessed 27 March 2019).

Steffen, W., Richardson, K., Rockstrom, J. et al. (2015) 'Planetary boundaries: Guiding human development on a changing planet', *Science*, **347** (6223), DOI: 10.1126/science.1259855.

Stern, N. (2006) *The Economics of Climate Change* (Stern Review), Cambridge: Cambridge University Press.

Stern, N. (2007) 'The economics of climate change', Talk given to the World Bank, Jakarta, 23/3/2007. see: https://slideplayer.com/slide/8948595/ (accessed 15 March 2019).

Sukhdev, P. (2013) 'Transforming the corporation into a driver of sustainability', in *State of the World 2013: Is Sustainability Still Possible?* ed. L. Starke, Washington: Island Press, pp. 143–153.

UNEP. (2011) *Recycling Rates of Metals: A Status Report*, Nairobi: United Nations Environment Programme.

Vitousek, P., Ehrlich, A. and Matson, P. (1986) 'Human appropriation of the products of photosynthesis', *BioScience*, **36** (6): 368–373.

Wackernagel, M. and Rees, W. (1996) *Our Ecological Footprint: Reducing Human Impact on the Earth*, Gabriola Island, BC, Canada: New Society Publishers.

Wallace-Wells. (2019) *The Uninhabitable Earth: Life After Warming*, New York: Tim Duggan Books/Penguin-Random House.

Wallich, H. C. (1972) 'Zero growth', *Newsweek*, 24 Jan, 1972.

Washington, H. (2013) *Human Dependence on Nature: How to help solve the Environmental Crisis*, London: Earthscan.

Washington, H. (2015) *Demystifying Sustainability: Towards Real Solutions*, London: Routledge.

Washington, H. (2017) *Positive Steps to a Steady State Economy*, Sydney: CASSE NSW. see: https://steadystatensw.files.wordpress.com/2017/06/posstepsroyal11ptjustheaderfinalju ne12thebooklowres.pdf (accessed 12 March 2019).

Washington, H. (2018a) 'Overshoot', in *Encyclopedia of the Anthropocene*, Volume IV, eds. D.A. DellaSala and M. I. Goldstein, UK: Elsevier, pp. 239–246.

Washington, H. (2018b) *A Sense of Wonder Towards Nature: Healing the World through Belonging*, London: Routledge.

Washington, H. (2019a) 'unpublished poem "Mind Trap"'.

Washington, H. (2019b) 'Ecosystem Services – A key step forward *or* anthropocentrism's "Trojan Horse in conservation"', in *Conservation: Integrating social and ecological justice*, H. Kopnina and H. Washington (eds.), New York: Springer, pp. 73–90.

Washington, H., Taylor, B., Kopnina, H., Cryer, P. and Piccolo, J. (2017) 'Why ecocentrism is the key pathway to sustainability', *The Ecological Citizen*, **1**: 35–41.

Wijkman, A. and Rockstrom, J. (2012) *Bankrupting Nature: Denying our Planetary Boundaries*, London: Routledge.

Wilkinson, R. and Pickett, K. (2010) *The Spirit Level: Why Equality is better for Everyone*, London: Penguin Books.

Wilson, E. O. (2003) *The Future of Life*, New York: Vintage Books.

Wilson, E. O. (2016) *Half-Earth: Our Planet's Fight for Life*, New York: Liveright/Norton.

Wilson, L. (2015) 'The carbon footprint of 5 diets compared', Shrink That Footprint website. see: http://shrinkthatfootprint.com/food-carbon-footprint-diet (accessed 12 March 2019).

WWF. (2018) *Living Planet Report - 2018: Aiming Higher*, World Wide Fund for Nature, M. Grooten and R. E. A. Almond (eds.). Gland, Switzerland: WWF.

WWF (n.d.) 'Threats: Deforestation, World Wide Fund for Nature', see: www.worldwil dlife.org/threats/deforestation# (accessed 12 March 2019).

2

A *HEALING* WORLDVIEW AND ETHICS

Of what
Does my madness consist?
What sets me apart
From the norm?
Well, to begin:
I believe in 'Nature'
I honour the wild
I listen to the land
I dare to say
That I revere it,
Respect it,
Indeed I love it!
How mad is that?

From 'Madness' (Washington 2013a)

SUMMARY

This chapter is about worldview and ethics, and how we cannot heal our world without transforming these. The key problem is that Western society (now globalised around the world) is dominated by anthropocentrism, amounting to a worldview of human supremacy that allocates no intrinsic value to nature. This shapes society's unsustainable interactions with the rest of life and is a major block to healing our world. We need an ecocentric worldview and an ecological ethics that see humans as part of nature, accepts that nature has intrinsic value and assumes a need to respect nonhuman nature and to uphold a 'duty of care' towards her. Anthropocentrism is deeply entrenched in society, government and academia. However, the good news is that if we change our worldview to ecocentrism, we *can* totally transform things. Indeed, this is one of the key changes needed if we are to heal our world. For the detail explaining this summary, please read on; however, if you just want to know the things *you can do* (five are listed) about this, then turn to the end of the chapter.

Introduction

What we need today is a healing worldview and ethics. The most dominant worldview today is Western society's anthropocentric modernism, now globalised around the world (Washington 2018a). Along with this worldview comes anthropocentric ethics, arguing essentially for human supremacy (Crist 2012). However, I argue here regarding anthropocentrism, modernism, neoliberalism, consumerism, resourcism – that none of these terms are actually *sane* views of the world if we want to live sustainably into the future. They are full of hubris, of a manic assumption that the only thing that matters is humanity, along with using more and more resources as fast as possible to increase the growth of GDP (see Chapter 3). Our current worldview is thus both insane and fundamentally *un*sustainable (Washington and Kopnina 2018), indeed it is the key cause of why our society and economy are broken, while nature is breaking. This chapter will explore why this is the case.

Albert Einstein (1950) once wisely observed:

> A human being is a part of the whole called by us universe, a part limited in time and space. He experiences himself, his thoughts and feelings as something separated from the rest, a kind of optical delusion of his consciousness. This delusion is a kind of prison for us, restricting us to our personal desires and to affection for a few persons nearest to us. Our task must be to free ourselves from this prison by widening our circle of compassion to embrace all living creatures and the whole of nature in its beauty. Nobody is able to achieve this completely but the striving for such achievement is in itself a part of the liberation and a foundation for inner security.

Einstein thus raised the key significance of how humanity *views* the world and the ethics that flow from this. D.H. Lawrence (1968: 504) has also wisely observed:

> This is what is the matter with us. We are bleeding at the roots, because we are cut off from the earth and sun and stars, and love is a grinning mockery, because, poor blossom, we plucked it from its stem on the tree of life, and expected it to keep on blooming in our civilised vase on the table.

Accordingly, we cannot properly talk about how to heal the world without talking about worldview, ethics and values. Note that many books on the environment and sustainability do not mention the first two, and only touch obliquely on 'values' (as discussed in Washington 2015). Some sustainability classics such a 'Cradle to Cradle' (Braungart and McDonough 2008: 11) even argue that: 'we won't solve problems if they are seen as ethical'. However, worldview and ethics are actually *central*, for if society does not question and address these, then it simply will not change. Without change, we will not heal the planet and reach a sustainable future. This chapter is adapted from chapter 4 in my book *A Sense of Wonder Towards Nature* (Washington 2018a). I discuss here worldview and ethics, anthropocentrism vs. ecocentrism, and the intrinsic value of nature. There is I believe a 'great ethical divide' between those who are anthropocentric and those who are ecocentric in society. Sadly, this has probably never been as great as it is today. Humanity is now more divorced from the natural world than ever before (Louv 2005). How did we get to this stage?

Worldview, ethics, values and ideologies

We read about the terms 'worldview', 'paradigm' and 'ideology'. Worldview is self-explanatory; it is how we view the world. A *paradigm* may be seen as related (but somewhat narrower) than a worldview. It is generally described as a 'distinct concept or thought pattern'. Paradigms are: 'like spectacles through which people perceive a situation' (Cavagnaro and Curiel 2012: 165). The lenses are formed by a person's principles, knowledge and experiences, and by the values of the surrounding society (Ibid.). An *ideology* is a set of conscious (and unconscious) ideas that constitute one's goals, expectations and actions. A related term is a 'meme' (Dawkins 1976), which is an idea, behaviour or style that spreads from person to person within a culture (i.e. an infectious idea). Harich (2012) argues that infection by 'false memes' (e.g. 'climate change is not real') is one of the key problems we face in Western society.

It is easy to switch from worldview to paradigm to ideology in the same discussion. However, arguably worldview is the wider view, followed by paradigm, with ideology the most specific. Finally, it should be noted that 'ethics' is about *right or wrong* conduct. This involves what we value. It is important to consider why our worldview or paradigm is crucial. Donella Meadows (1997: 84) explains this beautifully:

> The shared idea in the minds of society, the great unstated assumptions ... constitute that society's deepest set of beliefs about how the world works. ... Growth is good. Nature is a stock of resources to be converted to human purposes. ... Those are just a few of the paradigmatic assumptions of our culture, all of which utterly dumfound people of other cultures.

She concluded (Ibid.: 84):

> People who manage to intervene in systems at the level of a paradigm hit a leverage point that totally transforms systems. … In a single individual it can happen in a millisecond. All it takes is a click in the mind, a new way of seeing.

So our worldview is critical to how we view 'nature' and our relationship to her. The worldview society adopts of course strongly influences human culture and is the underpinning of our economy and political systems. Worldview may also either aid or block attempts to heal the world.

Anthropocentrism

Humans are not alone in the world, although Western society recently has behaved as if we were (Vilkka 1997). Humanity has become self-obsessed, we focus on ourselves, or at least the majority of us in Western society do. We focus on our society, our economy and only lastly on the nature that supports us. Even Green political parties fall into this trap (Curry 2011). But what exactly *is* anthropocentrism? It literally means 'human-centred' in regard to our ethics and what we value. The *Oxford English Dictionary* definition of anthropocentric is: 'Regarding humankind as the central or most important element of existence'.

In regard to this discussion, one should be aware of the 'Anthropocentric Fallacy' (Fox 1990; Eckersley 1992). It has been said that by being human: 'we can only be anthropocentric: we seek our own good, not what we suppose is nature's' (Lowenthal 1964: 40). However, this is actually just an ideological statement, not a truth (or rational argument). Taylor (1986: 67) points out that humans *can* take an animal's standpoint: 'without a trace of anthropocentrism', and make judgements of what is desirable from that standpoint. Human valuation is done by humans (as we all agree) but that makes it anthropo*genic* (carried out by humans) not anthropo*centric*. Our valuation clearly does not have to centre on ourselves. Just because humans can only perceive nature by 'human' senses, does not mean we cannot attribute intrinsic value to it (Fox 1990; Eckersley 1992). Humans are quite capable of cultivating a non-anthropocentric (i.e. ecocentric) consciousness (Fox 1990) and attributing intrinsic value to nature. Just because we are human does not in fact mean we have to be egotists and focus on ourselves. Western society overwhelmingly operates from strong anthropocentrism, which presupposes that nature is created from, or exists purely for, human purposes (Vilkka 1997). Strong anthropocentrism sees humanity as the only end in itself, where nature is never taken into moral account, as it is seen as without 'agency' – *dead* (Vetlesen 2015). Vilkka (1997: 102) concludes:

> By ignoring the significance of animals and nature, strong anthropocentrism is not even interested in real human well-being. It may be interested in the human economy and industrial development, but not the well-being of people or societies in their cultural, religious, aesthetic and environmental aspects.

There are other terms that are part of the anthropocentric lexicon. The first is *humanism*, where humans are seen as the most significant beings on Earth (Vilkka 1997). An anthropocentric attitude to nature is necessary to be a humanist. The basic feature of humanism is to defend the good of human beings (Ibid.). The 'Arrogance of Humanism' (Ehrenfeld 1978) is still abundantly alive in human practices. Humanism and humanitarianism are linked to anthropocentrism, affirming the human side of the nature/culture pair (Plumwood 2001). Plumwood argues that humanism has arguably helped us to lose touch with ourselves as beings that are also natural and have their roots in the Earth (Ibid.).

Crist (2012) argues that human supremacy has become so entrenched in society that the wondrous diversity of life is reduced down to just 'resources' or 'natural capital'. Seeing the beauty of the world as just 'things' (resources for our use) has become normal. Crist (2012: 145) argues that the concept of 'resources' has become: 'a gaping wound on the face of language' and engraved the delusion of human supremacy into 'common sense', science and politics. She believes that if we continue human supremacy, it will extinguish the possibility of: 'yet-to-be-imagined (sane, harmonious, beautiful) ways of being on Earth' (Ibid.). It should be emphasised that anthropocentrism is not about compassion for humanity; it is about *only* humanity having value and moral consideration in ethics. It also removes all compassion for nature from ethics.

The history of anthropocentrism

Many Indigenous societies had an ecocentric worldview (Knudtson and Suzuki 1992; Curry 2011). However, anthropocentrism has dominated Western society since at least the 16th century (Smith 1998). The universality and insidious aspect of anthropocentrism in modern culture has been attested to by many scholars. These include: Naess (1973), Ehrenfeld (1978), Godfrey-Smith (1979), Shepard (1982), Smith (1998), Taylor (2010), Curry (2011), Rolston (2012), Washington et al. (2017) and Kopnina et al. (2018). Plumwood (2002) and Vetlesen (2015) note that for at least four centuries we have regarded and treated nonhuman nature as something *dead*. There has also been a clear tendency for philosophers to focus primarily on the human mind. A central assumption of modernist Western moral thought has been that value can be ascribed to the nonhuman world only in so far as it is good: 'for the sake of the well-being of human beings' (Godfrey-Smith 1979: 310). The Western modernist attitude towards nature thus has a deep anthropocentric bias. Western culture has become ingrained with a doctrine of inherent human superiority, and this has become an unfounded dogma of our culture (Taylor 1986). It has been noted that if we conceive of nature as a 'machine', then that allows the human mind to retain a God-like position 'outside' of the world (Abram 1992). If the worldview of 'nature as a machine' rose to prominence in the 17th century (following the work of Descartes and Newton) due to its compatibility with a 'Divine

Creator', it remains in prominence today largely due to the deification of human powers that it promotes (Ibid.). Seeing nature as a machine, something less than ourselves, thus allows us to pretend we are something *more*.

Hay (2002) noted that there is currently something of a backlash within society against ecocentrism, and in favour of anthropocentrism, with greater emphasis on social justice and emergent democracy. This backlash demonstrates the insidious nature of anthropocentrism as an assumption in Western society, now sadly being disseminated around the world by globalisation. Anthropocentrism is also dominant in much of academia and government, and Washington et al. (2017) cite four examples of this, being ecosystem services, weak (and also strong) sustainability, education for sustainability and 'new' conservation. Indeed, anthropocentrism in academia may well be increasing. 'New conservation' (coming from the political Right) is a neoliberal approach in which conservation is undertaken primarily for human benefit, while 'critical social scientists' (coming from the political Left) see nonhuman nature as valuable only indirectly, insofar as it contributes to human well-being and social justice (as discussed by Kopnina et al. 2018). Both approaches to conservation are equally rooted in strong anthropocentrism, however, despite coming from different ends of the political spectrum (Ibid.).

The impracticality of anthropocentrism

Washington (2018a) notes that quite a few in academia insist that society does *not* have to change its worldview (e.g. Norton 1984). For example, Hatfield-Dodds et al. (2015: 49) state: 'Nor do we find that a shift in societal values is required' to become sustainable. They thus implicitly accept the anthropocentrism and modernism that have created the environmental crisis. We need to ask whether anthropocentrism is in fact *practical* in terms of society existing sustainably in the long term. The problem is that humanity has obligate dependence on nature to survive (Washington 2013b). For all our vaunted intelligence, technology and creativity, the food we eat, the water we drink, the air we breathe – all come to us from nature, which also cycles the nutrients in our crops and provides all the other myriad ecosystem services that society must have (MEA 2005; Wijkman and Rockstrom 2012). We need to live in the real ecological world that supports humanity. That means respecting ecological limits, many of which we have already exceeded (MEA 2005; Kumar 2010; Wijkman and Rockstrom 2012; Steffen et al. 2015; GFN 2019).

The well-being of nature supports us; hence we do not behave reasonably if we harm nature or decrease biodiversity (Vilkka 1997). Yet we are doing both at an accelerating rate, and it is neither reasonable nor in humanity's long-term interests (Wijkman and Rockstrom 2012; Washington 2013b). Given that anthropocentrism encourages us to live *un*sustainably by ignoring ecological limits (thus degrading the very ecosystems that support us) it is actually an impractical and unworkable ideology – a dead end. If we are interested in the long-term well-being of humanity, then we must be interested in the long-term well-being of nature. Thus, 'respect for nature' is a necessary and essential thing, yet anthropocentrism denies and rejects this, as nature is stereotyped as merely something to be used as humans see fit. Anthropocentrism does not respect nature; in fact it denigrates it (Washington 2018a).

The psychology of anthropocentrism: paranoia, fear and denial

The problem with anthropocentrism can be expressed by the phrase: 'We do not have the idea, the idea has us' (Wenz 2001: 238). The psychology of anthropocentrism is isolating, paranoid, fearful and aggressive, with a focus on an impossible: 'mantra of perpetual growth' (Rees 2008). It is solipsistic (extremely egotistical), believing that the world revolves about humanity. In effect we are meant to believe that life has evolved on Earth for 3.5 billion years, flowering into an amazingly diverse web of life – just to become the resource plaything of a species that has existed for perhaps 500,000 years (Washington 2018a). This arrogance and egotism isolates humanity from everything else, from the beauty and glory of the landscape, and from the wonder of the living world – from which we evolved. Plants and animals are not just 'resources' for us, they are our relatives, far and near, on the amazing odyssey of evolution we have all been part of (Leopold 1949). The closing off of humanity from the rest of life, and the debasement of the living world to become just 'things to be used', has placed humanity in a paranoid situation (Washington 2018a). Anthropocentrism encourages us to think we are the truly privileged 'top dog', the 'Masters of Nature'. Many seem to think that to stay at the top we must brutally overpower our enemy – being all other forms of life. Human mastery seeks to subjugate this unknown and scary natural 'other'. In reality we never 'conquered' nature, we just raided fossil fuels to cause the environmental crisis (Greer 2009). We stole fossil energy from the past, and (due to the consequent impacts) we are now robbing the future, making it unsustainable for those who follow us (Washington 2015).

Above all, anthropocentrism turns the wonder of the living world into an object of fear, even an object of hatred (Washington 2018a). It is a truism that humans tend to destroy what they fear. Fear is not a good aspect of the human psyche. Curry (2011) notes that anthropocentric modernism disenchants reality and inanimates nature. Fear separates a developing child from the full essential benefits of nature. Fear of traffic, crime, stranger-danger and of nature itself. As Louv (2005) concludes, we may end up teaching our kids that life is too risky, but also not real.

Humanity has a key flaw: *we tend to deny a reality we do not like*. This is discussed further in Chapter 3. Yet denial is the worst aspect of humanity, and today it has in effect become a pathology. Ignoring reality is not aiding humanity; hiding our heads in the sand is not a survival strategy, quite the opposite (Washington and Cook 2011). So in summary, anthropocentrism is a worldview that leads to paranoia, fear and denial, hence it is not a good thing for us humans. It doesn't tap into the best of humanity, where we accept our problems, cooperate and work together (which is how we lived successfully as tribal societies, Gowdy 1997). It doesn't spark our imagination and creativity; it doesn't assist our 'co-intelligence' to solve problems (Atlee 2003). Most importantly, anthropocentrism doesn't help us heal the world; in fact it is a key barrier to such healing.

Ecocentrism

Anthropocentrism regards humanity as the central element in the Universe; *ecocentrism* is instead focused on a nature-centred system of values and accepts that humanity is part of nature and must treat it with responsibility and respect. Ecocentrism finds intrinsic value in nature. It takes a much wider view of the world than does anthropocentrism, which sees the human species as more valuable than all other organisms. Ecocentrism is the broadest of worldviews, but there are related worldviews (that might be called 'intermediate varieties', Curry 2011: 57). Ecocentrism goes beyond *zoo*centrism (seeing value in animals) on account of explicitly including plants and the ecological contexts for organisms. It goes beyond *bio*centrism (ethics that sees inherent value to all living things) by including environmental systems as wholes, and their non-living aspects. Ecocentrism is the umbrella that includes zoocentrism and biocentrism, and all three of these worldviews value the nonhuman, with ecocentrism having the widest vision. Given that life relies on geology and geomorphology to sustain it, the broader term 'ecocentrism' seems the more inclusive concept and value and hence most appropriate (Washington et al. 2017). Curry (2011: 93) concludes that all dark or deep green ethics subscribe to ecocentrism, and to the position that the ecological community forms the ethical community. Curry (2011: 94) argues a final reason for giving ecocentrism 'pride of place' in ethics is the *urgency of its relevance*, which: 'seems matched only by the extent to which it has been ignored or disparaged'. Extraordinary as it may seem, feminism, anti-racism and socialism are almost as likely to ignore ecocentric ethics as neoliberalism (Ibid.). This demonstrates just how deeply embedded anthropocentrism is in society and academia, where even supposedly 'radical' and 'progressive' movements fail to foreground ecocentrism. Similarly, even parts of the conservation movement suffer from a growing tendency to anthropocentrism (Miller et al. 2014; Kopnina et al. 2018; Kopnina and Washington 2019). Hence the need for a major transformation of worldview and ethics.

Is ecocentrism anti-human?

Ecocentrism and belief in the intrinsic value of nature has been labelled 'anti-human' (Smith 2014), or as contrary to concerns for social justice. This claim is simply mistaken (Curry 2011). I agree with Stan Rowe (n.d.: 4):

Ecocentrism is not an argument that all organisms have equivalent value. It is not an anti-human argument nor a put-down of those seeking social justice. It does not deny that myriad important homocentric problems exist. But it stands aside from these smaller, short-term issues in order to consider Ecological Reality. Reflecting on the ecological status of all organisms, it comprehends the Ecosphere as a Being that transcends in importance any one single species, even the self-named sapient one.

Ecocentrists overwhelmingly support inter-human social justice; however, they also support inter-species justice, or *ecojustice*, for the nonhuman world (Baxter 2005; Washington et al. 2018). Just as environmental systems involve many interrelationships, environmental and social systems are entwined, and so social and ecological justice concerns are (and must be) as well (Washington 2015).

Intrinsic value free from human valuation

Nature and life are *inherently good*. That is a fundamental ethical conclusion of many environmental philosophers and scholars (Sweitzer 1949; Schopenhauer 1983; Vilkka 1997; Curry 2011; Rolston 2012; Vetlesen 2015; Weber 2016; Washington et al. 2017). Such an ecocentric ethical position lends support to con-servation efforts to protect the living world, while anthropocentrism fails to offer nature respect and hence makes conservation that much harder. To say nature is an 'inherent good' is to say it has *intrinsic value*, irrespective of whether humans are the ones valuing it. The idea that nature has 'intrinsic value', a right to exist for itself, irrespective of its use to humanity, was part of many Indigenous cultures (Knudtson and Suzuki 1992; Curry 2011). It was also resurrected by the writings of the Romantic poets such as Wordsworth and Coleridge. Later it was espoused by the visionaries Thoreau, Muir and Leopold (Oelschlaeger 1991) and later again by Rachel Carson (Washington 2018a). Vilkka (1997) notes however that most philosophers and scientists do not accept the language of intrinsic value in relation to the nonhuman world. She questions two common assumptions: 1) only humans have intrinsic value; and 2) humans are the origin of all values.

When discussing 'intrinsic value' one must be careful as to what one means (Washington 2015). The term carries a lot of philosophical baggage. It is compli-cated by a past history of attack by pragmatism, debates about objectivism and subjectivism, and by claims equating it to hedonism (Monist 1992). Much of the discussion is about the term 'intrinsic' itself, rather than whether nature as such has value. Several philosophers agonise about whether certain parts of nature have *more* intrinsic value than others (Ibid.). However, to skirt this can of worms, I use 'intrinsic value' here for the simple idea that the nonhuman world has value irre-spective of whether it is of use to humanity. Accordingly, we should respect it, indeed have a 'duty of care' towards it (Washington 2018a). Regarding intrinsic value, Rolston (1985: 30) argued: 'such values are difficult to bring into decisions; nevertheless, it does not follow that they ought to be ignored'.

There has been a process of 'ethical extensionism' (Vilkka 1997: 19) in regard to intrinsic value, where intrinsic value has extended from: 1) just humanity, 2) to sentient beings, 3) to all of life, 4) to ecosystems, 5) to geodiversity, 6) to the whole planet (Washington 2018a). Singer (1975) argues that only sentient beings are of intrinsic value. Several philosophers argue that *the value of life* is the fundamental principle in ecological ethics (Sweitzer 1949; Schopenhauer 1983; Vilkka 1997; Curry 2011). The well-being and flourishing of human and nonhuman life thus have intrinsic value, independent of the value of that life for human purposes. Richness and diversity of life-forms are thus values in themselves (Vilkka 1997). Johnson (1991) argues ecosystems (and the biosphere) have intrinsic value. This can be called 'biocentric holism' (Vilkka 1997), where ecosystems have *interests* and thus belong in the scope of morality. However, Johnson extends the moral arena to all 'living' things, but not to the non-living. People have been slower to extend intrinsic value to the non-living part of ecosystems known as 'geodiversity' (Washington 2018b), though Leopold (1949) does extend this to 'the land'. Hargrove (1992) and Gray (2013) argue that non-living nature such as geodiversity should also be seen as having intrinsic value. Planet Earth is itself increasingly being granted intrinsic value (and rights) by people (Berry 1988; Vilkka 1997; Curry 2011) and this is part of movements such as Earth jurisprudence (Cullinan 2003). I support the broadest ethical extension of intrinsic value to *all* of these six categories (Washington 2018a).

The intrinsic values and 'Rights of Nature' have been recognised in the 'World Conservation Strategy' (IUCN 1980) and in the Earth Charter (2000). The 'Millennium Ecosystem Assessment' (MEA 2005) and the UNEP project 'The Economics of Ecosystems and Biodiversity' (Kumar 2010) both acknowledge it (even if only in passing). A new positive development is that the UN has now established a 'Harmony with Nature' programme (www.harmonywith natureun.org).

Finally, regarding intrinsic value, Curry (2011: 140) notes that the intrinsic value of nature is ultimately an *inexhaustible mystery*: 'it cannot be fully explained, analysed or justified in terms of other concepts or values; otherwise it would not be "intrinsic"'. When I set foot in wilderness or old growth forest (or any intact natural area), the idea that only humanity has value seems bizarre and nonsensical. The wonder, harmony and beauty of such places is real, there to be easily discovered by all who listen and embrace empathy (Washington 2018a). Nature is the *generator of value*, having created so many wondrous and amazing things, including humanity. Through wonder, we engage with the beauty and unique value of different aspects of nature. The fall of rain in a storm, the rise of a full moon, the curving outline of an ancient tree, the light in an animal's eye. All of our evolutionary kin, plus the geodiversity that forms their homes, have value. How can they not? Only the blinkers and arrogant hubris of strong anthropocentrism cuts people off from a recognition of such intrinsic value, a value that most children understand (Ibid.). Time now for Western society to remove the blinkers and rediscover the intrinsic value of the rest of nature in order to heal the world.

BOX 2.1 HEALING EXCEEDING OUR POWERS? HOLMES ROLSTON III

We cannot heal the environmental crisis without confronting how we face an unprecedented hinge point in the 45 million centuries of life on Earth. Our so-called wise species, *Homo sapiens*, is jeopardizing the future of this wonderland home planet.

We have to think big – bigger than humans have ever thought before. We have to wonder and wander our way through what might almost be impossible for us. Solving our emergency is not in our genes. We have to act on a larger scale than any of which our ancestors dreamed – saving ourselves as Earthlings on a planet we are pushing toward collapse.

The promise and the peril is the same thing: our enormous powers, multiplied by our genius ten thousand times over what our great grandfathers had – from muscle and blood to nuclear energy. These powers escalate our desires. Our evolutionary past did not give us many genetic controls on our desires for goods that were in short supply. We love sweets and fats, of which in Pleistocene times humans could seldom get enough. But now we grow fat. When we can consume, we love it, amass goods, and overconsume. We have the same desires as our ancestors, but suddenly have acquired vast amounts of power and race to satisfy these desires. 'The faster a blind horse runs, the sooner it will perish'.

We congratulate ourselves. Humans are endowed with a remarkable mind, by far the most complex thing known, of virtually infinite complexity, capable of semantic and symbolic speech. We generate cumulative transmissible cultures, elaborating high orders of rational and emotional thought in science, philosophy, ethics, and religious faith. Humans alone ask who they are, where they are, and what they ought to do.

Such wonder can escalate the danger. Humans have gained vast powers for transforming their planet through agriculture, industry, and technology. By recent accounts human dominance is so extensive that Earth has entered a new age, the Anthropocene Epoch.

Enthusiasts advocate that humans ought to engineer Earth resourcefully for increasing human benefits, bringing ever-increasing human domination of the landscape. The Anthropocene is 'humanity's defining moment' (American Geosciences Institute). We are 'the God species' (Mark Lynas, National Geographic). The technosphere supersedes the biosphere.

Think again. Does this fulfill human destiny or display human arrogance, failing to embrace our home planet in care and wonder? Ought not this sole moral species do something less self-interested than value all the produce of an evolutionary ecosystem only for the benefits they bring? Such an attitude hardly seems biologically informed, much less ethically adequate.

Yes, the future will be unprecedented. We must learn to manage ourselves as much as the planet. True, the future holds advancing technology.

> But equally, we do not want to live a de-natured life, on a de-natured planet. A good planet is hard to find.
>
> A better hope is for a tapestry of cultural and natural values, not a trajectory further into the Anthropocene. Keep nature in symbiosis with humans. Our future ought to be the Semi-Anthropocene, kept basically natural and entered full of cares for both humans and nature on this marvelous home planet. Is this beyond our powers?

Conclusion

It may be a hard pill to swallow to accept that the dominant worldviews of Western society (anthropocentrism, modernism, consumerism, resourcism) are *all* both unsustainable and insane. These are the worldviews that have wounded the world – and which we need to abandon. Discussing worldview is thus critical to creating change to heal our world (Washington et al. 2017; Washington 2018a). A key worry is the scarring that anthropocentrism does to the minds of us humans. It is such an insidious and destructive ideology, one that maims humanity's sense of wonder, cutting us off from the rest of life (Washington 2018a). Recognising the problem of anthropocentrism can be made real by considering 'natural justice', where we should be just to the nonhuman world. How can all intrinsic value, all meaning, all moral standing, *just be limited to humans* (Ibid.)? Such a position reflects self-obsession, egotism and a madness (Shepard 1982) that focuses only on humans, and devalues and disenchants the rest of life. Justice that is only 'justice for humans' is not true justice. The rest of life, indeed the rest of nature as a whole (including geodiversity), also deserves justice (Washington et al. 2018). Anthropocentrism denies this, while ecocentrism acknowledges and supports it. Anthropocentrism is thus a great moral wrong, a key barrier to healing the world.

Ecocentrism does not mean humanity cannot make use of nature. What it does do is situate humanity as part of the web of life. Like all species, humanity needs food and shelter, it is just we require (or at least want) *more* than most species, via the medium of our numbers, our culture and our technology (Washington 2018a). Ecocentrism acknowledges our kinship with the rest of life, and implies a respect for nature and assumes a 'duty of care' to protect it. Yes, of course instrumental values of nature do exist, we do need to raise food, build houses, provide water, etc. Acknowledging the intrinsic value of nature means adopting the 'Land Ethic', where humanity is just a plain member of the living community (Leopold 1949). Human values and needs should thus not always override nature's values and needs. Respecting nature means not degrading it, not pushing ecosystems into collapse.

Living sustainably as 'human nature' (for humanity is part of nature) requires living respectfully with 'more-than-human' nature (Abram 1996). That means living respectfully within the Earth's ecological limits (see Chapters 1 and 3). It means accepting reciprocity – we have been gifted so much by nature, so we in turn should *give something back*, rather than endlessly 'taking'. Due to anthropocentrism, we have broken our covenant with the Earth, with life itself. We have abandoned restraint and, in effect, become a plague species (Washington 2018a). In so doing, we are destroying the natural world on which we ourselves depend. Ecocentrism is not new; it is a return to the 'wisdom of the elders' of many Indigenous cultures. It is a return to an ethic of living in harmony with the wondrous living world from which we evolved. Indeed, given our predicament, it is time for us to adopt a 'Nature First' assumption (Washington and Kopnina 2019).

The good news is that if we change our worldview to ecocentrism, we *can* totally transform systems (as Donella Meadows 1997 noted). It is a key leverage point for change, for healing. Moving to ecocentrism means respecting nature; it means feeling a duty to protect nature – as she is our kin. It means feeling a responsibility to live within ecological limits, or rather to return to these, as we have already far exceeded several limits (see Chapter 1). Changing our worldview from anthropocentrism to ecocentrism is thus one of the key (and most useful) steps any of us can take to heal the planet.

What can *I* do?

1) Adopt an ecocentric worldview and *talk about worldview and ethics*. Don't just assume everyone you meet operates from the same worldview and ethics. If people make statements about 'human nature' being naturally greedy, then challenge these. Society has not always lived by anthropocentric modernism and neoliberalism, and it does not need to stay buried in such ideologies. It is essential that we break free and value the rest of life, and assume a 'duty of care' towards the planet. We need to adopt a focus common in Indigenous cultures, one of caring and sharing (Knudtson and Suzuki 1992).

2) Help to educate your children about ecocentrism and ecological ethics. Ask them whether their pets, the old trees down the road, the sacred places they found while walking through nature – do these have value and a right to exist for themselves? If they say they do, then suggest they think about the ethics of what is happening with the environmental crisis and the need for positive (and eco-ethical) change.

3) Take your children into the 'more-than-human' world for walks and camps, and share your sense of wonder towards nature (see Chapter 9).

4) Support the creation of 'Wild Law', Earth jurisprudence and the 'Rights of Nature', where such rights should be built into legislation at all levels (Cullinan 2003, 2014). Support a 'Planetary Declaration on the Rights of Nature' (Higgins 2010).

5) Support 'justice for nature' or *ecojustice*. Also support 'ecodemocracy', where proxies are appointed in governance systems to represent nature (see www. ecodemocracy.net).

References

Abram, D. (1992) 'The mechanical and the organic: On the impact of metaphor in science', *Wild Earth*, **2** (2): 70–75.

Abram, D. (1996) *The Spell of the Sensuous*, New York: Vintage Books (Random House).

Atlee, T. (2003) *The Tao of Democracy: Using Co-Intelligence to Create a World that Works for All*, Manitoba, Canada: The Writers Collective.

Baxter, B. (2005) *A Theory of Ecological Justice*, New York: Routledge.

Berry, T. (1988) *The Dream of the Earth*, San Francisco: Sierra Club Books.

Braungart, M. and McDonough, W. (2008) *Cradle to Cradle: Remaking the Way We Make Things*, London: Vintage Books.

Cavagnaro, E. and Curiel, G. (2012) *The Three Levels of Sustainability*, Sheffield: Greenleaf Publishing.

Charter, E. (2000) 'The Earth Charter', Earth Charter Initiative, see: http://www.earthcharterinaction.org/content/pages/Read-the-Charter.html (accessed 11/4/13).

Crist, E. (2012). 'Abundant Earth and the population question', in *Life on the Brink: Environmentalists Confront Overpopulation*, eds. P. Cafaro and E. Crist. Georgia: University of Georgia Press, pp. 141–151.

Cullinan, C. (2003) *Wild Law: A Manifesto for Earth Justice*, Totnes, Devon: Green Books.

Cullinan, C. (2014) 'Governing people as members of the Earth community', in *State of the World 2014: Governing for Sustainability*, ed. L. Mastny. Washington: Island Press, pp. 72–81.

Curry, P. (2011). *Ecological Ethics: An Introduction*, 2nd ed. Cambridge: Polity Press.

Dawkins, R. (1976) *The Selfish Gene*, Oxford: Oxford University Press.

Eckersley, R. (1992) *Environmentalism and Political Theory: Toward an Ecocentric Approach*, London: UCL Press.

Ehrenfeld, D. (1978) *The Arrogance of Humanism*, New York: Oxford University Press.

Einstein, A. (1950) Quote from a letter of condolence to Norman Salit on March 4, 1950, as shown in the Albert Einstein Archives, 61–226, Hebrew University of Jerusalem, see: http://aefind.huji.ac.il/vufind1/Record/EAR000025410/TOC#tabnav (accessed 27 March 2019).

Fox, W. (1990) *Toward a Transpersonal Ecology: Developing New Foundations for Environmentalism*, Boston: Shambhala.

GFN. (2019) 'World footprint', Global Footprint Network, see: http://data.footprintnetwork.org/#/countryTrends?cn=5001&type=BCpc,EFCpc. (accessed 23 March 2019).

Godfrey-Smith, W. (1979) 'The value of wilderness', *Environmental Ethics*, **1**: 309–319.

Gowdy, J. (1997) *Limited Wants, Unlimited Means: A Reader on Hunter-Gatherer Economics and the Environment*, Washington: Island Press.

Gray, M. (2013). *Geodiversity: Valuing and Conserving Abiotic Nature*, 2nd ed. Hoboken: John Wiley & Sons.

Greer, J. M. (2009) *The Ecotechnic Future: Envisioning a Post-peak World*, Canada: New Society Publishers.

Hargrove, E. (1992) 'Weak anthropocentric intrinsic value', *The Monist*, **75** (2): 83–207.

Harich, J. (2012) 'The dueling loops of the political powerplace', Twink organization paper, May 5th 2012, see: http://www.thwink.org/sustain/articles/005/DuelingLoops_Paper.htm (accessed 19 February 2018).

Hatfield-Dodds, S. H., Schandl, H., Adams, P. D., et al. (2015) 'Australia is "free to choose" economic growth and falling environmental pressures', *Nature*, **525**: 49–53.

Hay, P. (2002) *Main Currents in Western Environmental Thought*, Sydney: UNSW Press.

Higgins, P. (2010) *Eradicating Ecocide: Laws and Governance to Prevent the Destruction of Our Planet*, London: Shepheard Walwyn Publishers Ltd.

IUCN. (1980) *World Conservation Strategy: Living Resource Conservation for Sustainable Development*, Switzerland: International Union for Conservation of Nature and Natural Resources.

Johnson, L. (1991) *A Morally Deep World: An Essay no Moral Significance and Environmental Ethics*, Cambridge: Cambridge University Press.

Knudtson, P. and Suzuki, D. (1992) *Wisdom of the Elders*, Sydney: Allen and Unwin.

Kopnina, H. and Washington, H. (2019) *Conservation: Integrating Social and Ecological Justice*, New York: Springer.

Kopnina, H., Washington, H., Gray, J. and Taylor, B. (2018) 'The "future of conservation" debate: Defending ecocentrism and the Nature Needs Half movement', *Biological Conservation*, **217**: 140–148.

Kumar, P. (2010) *The Economics of Ecosystems and Biodiversity: Ecological and Economic Foundations*, London: Earthscan.

Lawrence, D. H. (1968) 'A propos of lady Chatterley's lover', in *Phoenix II*. London: Heinemann, pp. 487–515.

Leopold, A. (1949) *A Sand County Almanac, with Essays on Conservation from Round River*, New York: Random House (1970 printing).

Louv, R. (2005) *Last Child in the Woods: Saving our Children from Nature-Deficit Disorder*, London: Atlantic Books.

Lowenthal, D. (1964) 'Is wilderness "paradise enow"? Images of nature in America', *Columbian University Forum*, **7**: 34–40.

MEA. (2005) *Living beyond Our Means: Natural Assets and Human Wellbeing, Statement from the Board, Millennium Ecosystem Assessment*, United Nations Environment Programme (UNEP), see: https://www.millenniumassessment.org/documents/document.429.aspx.pdf (accessed 19 February 2018).

Meadows, D. (1997) 'Places to intervene in a system: In increasing order of effectiveness', *Whole Earth*, **Winter** (91): 78–84.

Miller, B., Soule, M. and Terborgh, J. (2014) '"New conservation" or surrender to development?', *Animal Conservation*, **17** (6): 509–515.

Monist. (1992) 'The Intrinsic Value of Nature, Special Edition', *The Monist*, **75**: 2.

Naess, A. (1973) 'The shallow and the deep, long-range ecology movement: A summary', *Inquiry*, **16**: 95–100.

Norton, B. (1984) 'Environmental Ethics and Weak Anthropocentrism', *Environmental Ethics*, **6**: 131–148.

Oelschlaeger, M. (1991) *The Idea of Wilderness: From Prehistory to the Age of Ecology*, New Haven/London: Yale University Press.

Plumwood, V. (2001) 'Towards a Progressive Naturalism', *Capitalism, Nature, Socialism*, **Dec 2001**: 3–32.

Plumwood, V. (2002) 'Decolonising relationships with nature', *Philosophy, Activism, Nature*, **2**: 7–30.

Rees, W. (2008) 'Toward sustainability with justice: Are human nature and history on side?', in *Sustaining Life on Earth: Environmental and Human Health through Global Governance*, ed. C. Soskolne. New York: Lexington Books, pp. 81–94.

Rolston, H., III. (1985) 'Valuing wildlands', *Environmental Ethics*, **7**: 23–48.

Rolston, H., III. (2012) *A New Environmental Ethics: The Next Millennium for Life on Earth*, New York: Routledge.

Rowe, S. (n.d.) *Ecocentrism and Traditional Ecological Knowledge*, see: http://environment-ecology.com/ecology-writings/700-ecocentrism-and-traditional-ecological-knowledge.pdf (accessed 1 February 2017).

Schopenhauer, A. (1983) *The Will to Live: The Selected Writings of Arthur Schopenhauer*, ed. R. Taylor. New York: Ungar.

Shepard, P. (1982) *Nature and Madness*, London: University of Georgia Press.

Singer, P. (1975) *Animal Liberation*, New York: Avon Books.

Smith, M. J. (1998) *Ecologism: Towards Ecological Citizenship*, Buckingham: Open University Press.

Smith, W. (2014) *The War on Humans*, Seattle: Discovery Institute Press.

Steffen, W., Richardson, K., Rockström, J. et al. (2015) 'Planetary boundaries: Guiding human development on a changing planet', *Science*, **347** (6223), doi: 10.1126/science.1259855.

Sweitzer, A. (1949) *The Philosophy of Civilisation*, New York: Macmillan.

Taylor, B. (2010) *Dark Green Religion: Nature Spirituality and the Planetary Future*, Berkeley: University of California Press.

Taylor, P. (1986) *Respect for Nature: A Theory of Environmental Ethics*, Princeton: Princeton University Press.

Vetlesen, A. (2015) *The Denial of Nature: Environmental Philosophy in the Era of Global Capitalism*, London: Routledge.

Vilkka, L. (1997) *The Intrinsic Value of Nature*, Amsterdam: Rodolpi.

Washington, H. (2013a) *Poems from the Centre of the World*, lulu.com, see: http://www.lulu.com/shop/haydn-washington/poems-from-the-centre-of-the-world/paperback/product-21255751.html.

Washington, H. (2013b) *Human Dependence on Nature: How to Help Solve the Environmental Crisis*, London: Earthscan.

Washington, H. (2015) *Demystifying Sustainability: Towards Real Solutions*, London: Routledge.

Washington, H. (2018a) *A Sense of Wonder Towards Nature: Healing the World through Belonging*, London: Routledge.

Washington, H. (2018b) 'The intrinsic value of geodiversity', *The Ecological Citizen*, **1** (2): 137, epub-007, see: http://www.ecologicalcitizen.net/article.php?t=intrinsic-value-geodiversity.

Washington, H., Chapron, G., Kopnina, H., Curry, P., Gray, J. and Piccolo, J. (2018) 'Foregrounding ecojustice in conservation', *Biological Conservation*, **228**: 367–374. https://doi.org/10.1016/j.biocon.2018.09.011.

Washington, H. and Cook, J. (2011) *Climate Change Denial: Heads in the Sand*, London: Earthscan.

Washington, H. and Kopnina, H. (2018) 'The insanity of endless growth', *The Ecological Citizen*, **2** (1): 57–63.

Washington, H. and Kopnina, H. (2019) 'Conclusion – A just world for life?', in *Conservation: Integrating Social and Ecological Justice*, eds. H. Kopnina and H. Washington. New York: Springer, pp. 219–228.

Washington, H., Taylor, B., Kopnina, H., Cryer, P. and Piccolo, J. (2017) 'Why ecocentrism is the key pathway to sustainability', *The Ecological Citizen*, **1**: 35–41.

Weber, A. (2016) *The Biology of Wonder: Aliveness, Feeling, and the Metamorphosis of Science*, Gabriolia Island, BC, Canada: New Society Publishers.

Wenz, P. (2001) *Environmental Ethics Today*, Oxford: Oxford University Press.

Wijkman, A. and Rockstrom, J. (2012) *Bankrupting Nature: Denying our Planetary Boundaries*, London: Routledge.

3

REALITY, LIMITS, DENIAL AND GROWTHISM

Am I sleep-walking,
Towards the abyss?
It is true I feel unsettled,
I can see something,
From the corner of my eye …
But for now,
I will pretend it is not there,
And maybe it will go away?
I really don't like to worry.
I will remain content -
Secure in my delusion.

From 'Denial' (Washington 2013a)

SUMMARY

An essential requirement (if we are to heal the world) is to accept the reality of our predicament. That means we have to accept ecological limits. However, as Chapter 1 has shown, society has failed to do so. The key problem is that much of humanity tends to *deny* problems it would rather not face. This chapter discusses the various reasons why this is the case. It then discusses how we can 'break the denial dam' that holds back healing. The chapter then considers the idea embedded in Western society that 'resources are infinite', and shows why this is not (and cannot) be true. It also looks at the limits to nature's gifts to society, limits we have exceeded (and continue to do so). The chapter concludes by discussing a key problem wounding the world – the cult of 'growthism' or 'growthmania', which argues that on a finite planet we can continue to grow physically forever (no matter what harm this causes). For the detail explaining this summary, please read on; however, if you just want to know the things *you can do* (seven are listed) about this, then turn to the end of the chapter.

Introduction

Western society (now globalised around the world) has a fundamental problem – it refuses to accept reality and to respect the ecological and physical limits of our world. In fact it calls defeatist and 'doom-mongering' anybody who point outs we live on a finite planet, where our society is now well *beyond* ecological limits (Washington 2015). Even if people accept the reality of overshoot, as Chapter 6 shows, many people will still assume a rampant techno-optimism that asserts that our technology is so good we can bypass or overcome all limits. This is (in part) due to the incredible development of technology that has occurred in our lifetimes. In part it seems due to a belief in a *Star Trek* future, where we exploit the moon and asteroids, and later find one planet after another to plunder. In this vision we would leave the exhausted Earth and effectively become a galactic plague – finding planets, consuming their resources and moving on. This vision is part of *growthism* or *growth-mania*, the idea that we must grow forever, always expanding, never stopping (Daly 1991). However, as writer Edward Abbey (1977) has rightly pointed out, growth for the sake of growth is the ideology of the cancer cell.

This chapter will consider why humanity has a problem accepting reality – to such an extent that we commonly deny a reality we don't like. This means we can deny not only the problem – but also the solutions to that problem (Washington 2018a). However, as we will discuss, one *can* break the denial dam. We will also revisit biophysical limits, which come in various forms, such as limits of natural objects and minerals (anthropocentrically often called renewable and non-renewable 'resources') and limits to the gifts that nature provides society (and which anthropocentrically are often called 'ecosystem services', Washington 2019).

Problems with reality

The questioning of 'the real' serves merely to distance humanity from nature, and to hide problems that can hurt us. Nature is *real*, our ecological dependency on nature is real and human actions really are degrading the ecosystems that support real human societies (Cardinale et al. 2012; Washington 2013b). Postmodernists often speak of 'co-creating reality' (Heron 1992 in Reason and Torbert 2001: 13; Burr 2003), when actually there is a 'real' reality (one worthy of respect), which humans interpret differently. Some postmodernist academics argue that:

> [i]t is now difficult to sustain a position of 'naïve realism'. In scholarly circles it is difficult to suggest that the world exists outside our construction of it.
>
> *(Reason and Torbert 2001: 4)*

Thankfully, some scholars question such a polemic (e.g. Butler 2002; Curry 2011; Rolston 2012; Washington 2018b). However, if nature is not 'real' to us, then we are unlikely to feel a responsibility towards her, accord her ethical status, or think much about healing our world (for the world would become just a non-entity irrelevant to human desires). The apparent postmodernist loss of contact with the 'real' concerns many authors (e.g. Lopez 1986; Barry 1994; Bryant 1995; Gare 1995; Soule 1995; Plumwood 2003; Vetlesen 2015; Washington 2018b). It has been pointed out that: 'the landscape is not inert, and it is precisely because it is alive that it eventually contradicts the imposition of a reality that does not derive from it' (Lopez 1986: 277–8). Deconstructionists use 'nihilistic monism' to deny nature's reality and claim that wilderness is illusory (Soule 1995). It has also been argued that postmodernism involves a loss of contact with any reality beyond language and texts (Gare 1995). The problem with questioning reality is that: 'people have been deprived of the fixed reference points by which they previously oriented themselves … They have been "de-natured"' (Ibid.: 27). Others note that: 'the ontological existence of nature-in-itself is an indisputable fact' (Barry 1994: 391) and point out that there is a real physical world which exists, whether or not humans are aware of it. We perceive the world, and that perception is shaped by our senses and culture. We thus 'construct' our own *perception* of the world and nature, but: 'that construction does not of itself alter the physical world. It only alters what each of us thinks it is' (Ibid.: 391). It has rightly been noted that:

> If we turn our regard for nature more and more into clever philosophical word games, if we begin to think that we are intellectually creating nature rather than physically participating in it, we are in danger of losing sight of the real wolves being shot by real bullets from real aeroplanes, or real trees being clearcut, of real streams being polluted by real factories.
>
> *(Bryant 1995)*

We cannot solve the environmental crisis unless we face the reality of our predicament. Paul Dayton of Scripps Institution of Oceanography (in Louv 2005: 161) notes: 'Reality is the final authority; reality is what's going on out there, not what's in your mind or on your computer screen'. However, when challenged with the reality of the environmental crisis, this is still angrily denied by many anthropocentrists (Washington 2018a). Rees (2019: 133) notes that: 'Many people are startled to learn that most of what they believe to be true, most of what they think they know, is literally *made up*'. Vetlesen (2015) asks why anti-realist views about nature enjoy such currency today. He notes this is not only in intellectual trends 'a la postmodernism', but in popular culture as well. He concludes (p. 28) that: 'the more sophisticated and the more virtual technology you use in your daily life dealings with the world, the easier it may be to adopt a constructivist, anti-realist world view'. Hence why Louv (2011) concludes that the more one uses virtual technology, the more one needs to get out into real nature to counteract this.

At all levels, whether the big picture of climate change and species extinction, or the local level about whether yet another area of native vegetation should be cleared for development – we need to accept the reality of our predicament. We live in a world of ecocide, where we are pushing the rest of life on Earth into extinction by continuously expanding our activities. We don't have to do this if we accept reality, the problems are known – as are the solutions – and have been at least since Catton (1982) wrote his landmark book *Overshoot*. These have been restated again and again, such as by Ehrlich and Ehrlich (1991), Berry (1999), Soskolne (2008) and Brown (2011). If we accept reality we break the denial dam, then we can get on with the solutions that are urgently needed to heal the world.

The relationship between optimism, pessimism and realism is important, and this is discussed in Chapter 10.

Why do we deny reality?

All of the environmental problems described in this book are real. All of them have also been *denied*. Denied, not just by the odd crank, but by the majority of governments and 'we the people' over many decades (Oreskes and Conway 2010; Washington and Cook 2011). Not only does humanity en masse deny the environmental crisis, we tend to deny that we deny it. Psychologist R.D. Laing (1971) notes it is as if each of us were hypnotised twice: firstly into accepting pseudo-reality as reality, and secondly into believing we were not hypnotised. To heal the planet we need to grasp how big a problem denial actually is. As Heidegger (1955) noted, society seems to be: 'in flight from thinking'. As a society overall, we continue to act as if there is no environmental crisis, no matter what the science says. As Rees (2010) concludes:

> Modern society has been paralyzed by deep-seated cognitive dissonance, collective denial, and political inertia in dealing with the unsustainability conundrum.

More recently, our leaders seem to speak of becoming 'sustainable', while at the same time denying the need for any major change. Hence it remains common for many to speak of a 'sustainable development' that is actually based on endless growth – the key driver of the environmental crisis (Washington 2015). How is it possible for civilisations to be blind towards the grave approaching threats to their security, even when available evidence is accumulating about these threats?

BOX 3.1 THE MOST POWERFUL THING WE CAN DO TO FIGHT CLIMATE CHANGE IS BREAK CLIMATE SILENCE, JOHN COOK

The science is clear on how to avoid the worst impacts of climate change. We (as in all of human society) need to reduce our greenhouse gas emissions to zero in this century. This involves completely transitioning our society from fossil-fuel based energy to renewable energy. What a huge, intimidating, global task! What can we as individuals possibly do to help solve this crisis?

There are many actions we can take to help solve climate change. We can change how we live, reduce our carbon footprint, and live cleaner, more efficient lives. These small actions contribute to reducing the amount of energy consumed. But actions to reduce our personal carbon footprint have a limited impact on driving the much-needed transition to renewable energy. To cause that transition, we need social and political momentum.

Consequently, I would argue that the most impactful action we can do to solve the environmental crisis is to simply open our mouths and talk about climate change. It is an unfortunate statistic that most of the public who are concerned about global warming fail to talk about it with their friends and family. The main reason for this self-silencing is pluralistic ignorance – the misconception that not many people agree with us. The climate-concerned are ignorant of the fact that they're a plurality – most of the public are on board with the science of climate change. But we erroneously think others disagree with us, and thus we fail to bring up the subject. Our silence unfortunately reinforces the misconception that few people care about climate change, adding to a reinforcing spiral of silence.

By talking about the issue with our friends and family, we build awareness and social momentum for climate action. If enough of us explain to our elected officials that we care about the climate change issue and vote accordingly, politicians will get the message that failing to act with mean-ingful climate policy runs the risk of them losing their job (which is the only thing we know for sure they care about).

We still need to change our personal lives and adopt pro-environmental behavior. Walking the walk makes our talk more powerful. But ultimately, the societal-level transition needed to mitigate climate change will only come if enough people voice their concern about this issue. We can end the spiral of silence by breaking climate silence.

For more information: www.skepticalscience.com

Sadly, denial is as old as humanity, and possibly nobody is free from it (Zerubavel 2006). Poet T.S. Eliot (1936) in his poem *Burnt Norton* notes that: 'Human kind … cannot bear very much reality'. We proceed often in a cultural trance of denial, which Lifton and Falk (1982) referred to as 'psychic numbing', where people and societies block awareness of issues too painful to comprehend. This human incapacity to hear bad news makes it hard to solve the environmental crisis (Sukhdev 2010) or to heal our world. Indeed, Harich (2010: 64) speaks of denial as 'manipulative deception', and suggests that as our currently abysmally low ability to detect this rises, then: 'deception effectiveness will start to fall'. People will then start to see through fallacious beliefs such as endless growth.

The organised denial industry has played a major part in slowing or stopping action on environmental problems (especially climate change), holding back moves towards any real sustainability (Washington 2015). Oreskes and Conway (2010) detail the support conservative think tanks have given to denial and ask: 'What is going on?' The link that united the tobacco industry, conservative think tanks and a group of denial scientists, is that they were implacably opposed to *regulation*. They saw regulation as the slippery slope to socialism. They felt that concern about environmental problems was questioning the ideology of laissez-faire economics and free market fundamentalism (Washington and Cook 2011). These conservative bodies equate the free market with liberty, so for them:

> Accepting that by-products of industrial civilization were irreparably damaging the global environment was to accept the reality of market failure. It was to acknowledge the limits of free market capitalism ... science was starting to show that certain kinds of liberties are not sustainable – like the liberty to pollute.
>
> *(Oreskes and Conway 2010: 238–9)*

So if science impacts on conservative organisations' view of 'liberty', if it shows that regulation of pollution is needed, then science has to be opposed and denied. The basis for much denial is thus *ideological*, where science and the environmental crisis are denied due to a conservative ideological hatred of regulation affecting the free market.

Denial (at the society level) leads to worsening problems, conflict, starvation and disease, and in the end to the collapse of civilisation. This has happened many times before with past civilisations (Diamond 2005). Denial turns off much of our human intelligence, imagination, creativity and ethics. Yet many people (indeed also some in academia) still deny both the environmental crisis and the need to change society's worldview. Those of us who dare to point this out (and various environmental scholars have done this for several decades) are ridiculed, called 'prophets of doom' or even 'antihuman'. This happened with Malthus (re his 1798 essay), it happened with the team who wrote the 'Limits to Growth' (Meadows et al. 1972), it happened with Paul and Anne Ehrlich, and with David Suzuki, and their many eminent books on the environmental crisis and its solutions (e.g. Ehrlich and Ehrlich 1991; Suzuki 2010). And it continues to happen today, most strongly in regard to the climate crisis (Hansen 2009; Oreskes and Conway 2010; Washington and Cook 2011). Environmental scholars who dare to point this out still face a barrage of criticism and denial, sometimes even leading to the loss of their jobs in academia.

So why do we deny reality? Rees (2008) argues that in times of stress the brain's 'reptilian brain stem' (amygdala) will override the rational cortex, so we do stupid things. Thus our dedication to growth comes from these survival instincts. Rees (2008: 91) concludes that if we are to survive, we must assert our capacity for 'consciousness, reasoned deliberation and willpower' to: 're-write the "myths we live by" and articulate the necessary conditions for sustainability'. Rees (2010) expands on what he calls the 'triune' brain, where the reptilian brainstem and limbic system assume dominance over the rational neocortex. He raises the worrying issue (p. 7) that: 'Intelligence and reason may not be the primary determinants of human behaviour at any social scale'.

Do we let denial prosper? The answer is clearly 'yes'; society does so, and it does this through various mechanisms, as summarised by Washington (2015) in the following subsections.

Fear of change

Of course many of us are not in denial, but it seems that many people and governments are. One reason why many of us don't take action is the 'fear of change'. When struggling with the trauma of change, segments of society can turn away from reality in favour of a more comfortable lie. The fear of change is related to denial and is a strong trait in humanity. Indeed, it is the hallmark of conservatism (though some prefer to call this neoliberalism or radical liberalism, Giddens 1994). Many conservatives just don't 'trust' scientists (and certainly not environmentalists) as they think they are too 'liberal' (and hence suspect). Conservatives see environmental regulation as threatening the core elements of conservatism, such as the primacy of individual freedom, private property rights, laissez faire government and promotion of free enterprise (McCright and Dunlap 2000).

Failure in worldview and ethics

Another reason why denial prospers is a failure in worldview, ethics and values. There is little discussion of worldview or ethics in modern society (see Chapter 2). It is a truism that if you don't know where you are 'coming from' you don't know where you are going. If we don't strongly value the natural world (or even society) that is in crisis and threatened, then we probably won't feel the urgency to heal it.

Fixation on economics and society

Our society has a fixation on economics and society. Our society looks firstly to economics and secondarily to social issues (Washington 2015). The intelligentsia in our society has mainly focused on social justice, and 'ecojustice' is rarely given equal prominence, or even any consideration at all (Washington et al. 2018). Under these circumstances it is hard for many people to feel a sense of urgency, as ecological degradation is just not 'real' to them. This is partly due to the fact that Western society is now more isolated from the natural world than ever before (Louv 2005).

Ignorance of ecology and exponential growth

Another reason why we let denial prosper is our lack of an 'ecological ground-ing'. Most people don't understand how the world works in terms of ecosys-tems. The ecological grounding of our society has improved since the early 1990s, but many decision-makers are still woefully ignorant. Some of them still think environmental issues are just about 'tree hugging'. For this reason they feel no urgency to act on environmental issues. Another reason why people don't feel an urgency to act is our inability to really understand 'exponential growth' (Brown 1978). Many environmental problems are escalating at expo-nential rates, yet humans generally do not think exponentially. Failure to under-stand exponential growth leads to a failure to act urgently on environmental problems and aids denial.

Gambling on the future

Another aspect that delays us taking action is gambling on the future. People amazingly tend to discount the future (Giddens 2009). Many of us are gamblers at heart. If you are a gambler then you may not feel any sense of urgency, as you will always be gambling that things will turn out fine.

The media and communication

Another aspect that helps denial to prosper is the media, which Ehrlich and Orn-stein (2010) note has largely become a 'disinformation service'. The media poorly communicates environmental science. It also loves controversy. This is often justi-fied as being 'balanced' reporting, when in fact it shows major bias. A situation where the judgement of the vast majority of the scientific community is given equal space with denial advocates is anything but 'balanced'. In fact 'balance' has become a form of bias in favour of extreme minority views (Oreskes and Conway 2010).

Breaking the denial dam

Denial allows people to ignore all facts, all ethics and even all solutions. Everything I have learned from reading hundreds of books and papers, and working for over four decades in the environmental field, has confirmed this. However, we *can* (per-sonally and as a society) break the denial dam, accept reality and change things for the better. We must accept that denial is part of humanity; it has probably been with us since we first evolved. Indeed Sharot (2011) confirms humanity has an 'optimism bias' where we always think things will turn out better than they do.

We should understand there are those who operate from greed and deliberately create a denial industry to confuse people. This is immoral and destructive (Washington and Cook 2011). Similarly, there are the denial spin-doctors in government and business who seek to use 'weasel words' and fool the public that they are taking meaningful action, when often they just continue 'business-as-usual' (Ibid.). However, there is also denial in 'we the people'. We let ourselves be duped and conned, we let our consciences be massaged and we let our desire for the 'safe and easy life' blot out unpleasant realities (Ibid.). Hence we often delude ourselves. It is time to wake up and break the denial dam.

How big a problem is denial in terms of healing the world? Like worldview, denial is central to this. The environmental crisis is solvable; however, you don't solve a problem you deny exists. The solutions exist (see Chapter 11), but the solutions get harder the longer we deny the need for change. We could have acted decades ago when Catton (1982) wrote *Overshoot* – and would now have been well on the path to healing the world. Instead, we denied we had a problem and deluded ourselves. Yet humanity *has* solved major problems, we acted on acid rain, we acted on CFCs (causing the hole in the ozone layer), so we can also act for real sustainability (Washington 2015). Humanity (when it chooses) is intelligent and creative and can solve problems. The key issue now is to make that choice and break the denial dam. Denial and healing the world are mutually exclusive; if you have one then you don't have the other.

'Resources' are infinite only in our minds

I discuss here the commonly used term 'resources', even though the term has rightly been called a: 'gaping wound on the face of language' (Crist 2012: 145), one which has led to a distorted (and destructive) view of the world. The rocks and ores of the Earth are called 'non-renewable resources', while the living world (e.g. forests, fish, etc.) are called 'renewable resources'. In one stroke the use of the term 'resources' assumes that all biodiversity and geodiversity are just things for *our* use, without any intrinsic value or right to exist for themselves. It also assumes humanity 'owns' all biodiversity and geodiversity and the whole world (see Chapter 8). However, when talking about the minerals used by society, it is virtually impossible to avoid the use of the (limited and biased) term 'resources'; hence I make use of it here (though we should never forget its bias). To go back to basics, the Earth is physically *finite*, hence there is only a certain amount of 'stuff' in the Earth, being the Earth's crust, mantle, and inner and outer cores. All our minerals come from the Earth's continental crust (or atmosphere for nitrogen) or sea bed (e.g. manganese nodules). It is thus physically impossible to forever grow the amount of materials we extract from the Earth; indeed it is impossible to continue for very long (on any scale longer than a few decades) at our existing high extraction levels.

Daly (1991: 108) points out that an assumption of neoclassical economics is that 'resources are infinite', that when one dries up there will 'always be another', and technology will always find cheap ways to exploit this (= techno-centrism). As long as neoclassical economics continues to ignore the absolute dimension of scarcity, then it has: 'uncoupled itself from the world and become irrelevant' (Daly 1991: 108). Mining of ores and minerals grew 27-fold in the 20th century (UNEP 2011). Indeed, in the year 2000 the annual anthropogenic sediment flux from coal, metals, minerals, cement and aggregates production for that year is estimated to have been about 81 billion tonnes, six times the natural sediment flow of rivers (Cooper et al. 2018). This is even larger than the estimated annual loss of soil due to agriculture of 75 billion tonnes (Brown 2011). However, some economists have denied the relevance of matter and energy. For example, Gilder (1981) spoke of overcoming the 'materialistic fallacy', the illusion that resources and capital are things which can run out, rather than products of the human will and imagination, which he argues in freedom are inexhaustible. Most economists of course do actually recognise that production and consumption have physical dimensions, but most have also largely remained silent about openly stating this (Daly 1991).

Daly (1991: 283) shows there is a neoclassical proposition that human-made capital is a 'near perfect substitute' for natural resources (as noted by Nordhaus and Tobin 1973; Stiglitz 1979). There is thus the absurd implication that: 'we could build the same house with one tenth of the lumber if only we had more saws or hammers' (Daly 1991: 283). There thus exists a widespread modern neoclassical economic view that human capital can substitute for land or nature (= weak sustainability), and hence the goal of increasing capital can proceed without any attention to what is physically happening to the land (Ibid.: 110). This is an important driver of the environmental crisis. Yet this substitution argument is often cited in order to argue that natural resources scarcity need not limit growth (Ibid.). Economist Julian Simon (1981: 207) has stated about substitution: 'you see, in the end copper and oil come out of our minds. That's really where they are'. Such views are still influential and are rarely criticised by mainstream economists even now (Daly and Cobb 1994: 109). Indeed, even in 2017, a group of geologists (Arndt et al. 2017: vi) argued:

> We reach the fundamental conclusion that mineral supplies will be maintained at levels that satisfy demand well into the future, and that any peak of production of individual commodities will probably result from declining demand caused by technological or societal developments, rather than exhaustion of the resource.

I note the rubberiness of the phrase 'well into the future', when one would hope that an ecologically sustainable human society would still exist in tens of thousands of years. Similarly, the 'Ecomodernist Manifesto' (EM 2015: 10) maintains:

To the degree to which there are fixed physical boundaries to human consumption, they are so theoretical as to be functionally irrelevant.

A different perspective comes from mining expert Simon (Michaux 2014: 82) who believes we are near to confronting 'peak mining' and argues:

Everything we need and want to operate in our current way of life is drawn from non-renewable natural resources in a finite system. Most of those natural resources that we need are depleting or will do so soon. Conversely, demand for everything we need and want is expanding fast in the name of economic growth (and increasing population). When these trends meet, there will come a point where the way in which we do things will fundamentally change.

The above shows how polarised things have become in regard to the debate about the increasing use of mineral resources. Despite the obvious reality that 'on a finite planet you cannot grow physically forever' there is still major denial that (at some future point) some resources will become vanishingly scarce. The current dominant approach is determined by neoclassical economics and neo-liberalism, ideologies of endless growth that seek an endless growth in extraction of resources. Classical economic theory developed at a time when the environment was considered an 'infinite resource', but the scale of our economy is now vastly larger (Daly 1991: 185). Daly argues the concept of *throughput* should now displace the 'circular flow' model in economic theory. Daly warns us not to get carried away with the human mind being an 'ultimate resource' that can generate endless growth, for the mind is not independent of the body. Before leaving resource use, we should discuss *Zeno's Paradox* of 'Achilles and the Tortoise'. Achilles gives the tortoise a head start, but whenever Achilles reaches somewhere the tortoise has been, he still has farther to go. Because there are an infinite number of points Achilles must reach (where the tortoise has already been) – he can never overtake the tortoise! The paradox actually confuses an infinity of subdivisions of a finite distance with an infinity of distance. Yet this is what Simon (1981) similarly does with resources. He confuses an infinity of possible boundary lines between 'copper' and 'non-copper' with an infinite amount of copper. There is (after all) only a finite amount of copper on Earth. The flawed logic of Zeno's paradox, however, is still commonly argued to suggest resources have no limits (Daly and Cobb 1994).

So what are the limits to the Earth's minerals? As shown above, you will get different responses from different people who hold different worldviews. Those wedded to endless growth economics will still pretend any limits to resources are far away. Those committed to a more realistic assessment of mining (and its impacts) will argue for a major reduction in mining and other resource use, given the impact of such resource use accelerates ecocide. However, going back to first principles, there is only so much of any mineral (at various ore grades) on Earth. The richer the ore the less energy needed to extract the mineral; the poorer the ore the higher the energy needed. Given that oil itself as a resource is close to (or has passed) its peak where half of the total resource has been used (Almeida and Silva 2009) and we should not use coal for climate reasons (IPCC 2018), and should not use nuclear power for many reasons (see Chapter 6) – the age of 'easy energy' for mining has passed. Recycling should of course take place, but this is also never 100% efficient, as some material is always lost (or more correctly becomes too diluted to use economically). Thus a growing human population wanting more and more scarce elements (e.g. neodymium for magnets in wind turbines) will mean the resource becomes limiting. The technology we use, and a 'cradle to cradle' approach (Braungart and McDonough 2008), can delay peaks in such resources, but not avert them, especially if society is locked into an endless growth mantra. Hence various minerals will become harder and harder to extract, requiring more and more energy to do so (Michaux 2014). Examples are phosphorus for fertilisers (expected to peak in a few decades), copper for cables, neodymium for magnets (for wind turbines). Limits for minerals can thus be delayed but not ignored.

Accordingly, our high and growing use of what are called non-renewable resources must be changed. This means strategies such as 'Factor 5', where you use only 20% of energy and materials as we do currently (von Wiezsäcker et al. 2009) or even 'Factor 10' where you use only 10% (www.factor10-institute.org). However, it also means stabilising, and then reducing, population as swiftly as we can via a humane strategy (see Chapter 4). It also means we must relearn what our grandparents and great-grandparents knew – *thriftiness* (see Chapter 5). We must reinvent consumer culture (Assadourian 2013) to consume much less than today. Meadows et al. (2004: 54) summarise Daly (1990) arguing that:

> For a non-renewable resource (fossil fuel, high grade mineral ore, fossil groundwater, etc), the sustainable rate of use can be no greater than the rate at which a renewable resource (used sustainably) can be substituted for it.

Daly (1991) suggests 'depletion quotas', which operate by restricting the supply of a resource. The first stage would be a government auction of rights to purchase a certain amount of a resource. The total amount of purchase rights to be auctioned in a given year would be determined by legislative decision. However, while some governments have resource taxes, none seem to have adopted the depletion quota idea. The key problem is that most governments still want to sell *more* minerals, more iron, copper, lithium, phosphorus – not reduce these to conserve them in the long term. There is no conception of, let alone commitment to, *scarcity*, that many of the minerals we madly mine now are, on any timescale longer than 100 years – quite scarce. Yet we hope human civilization will exist for many thousands (indeed tens of thousands) of years. There seems to be a mental block of 'evermoreism' (Boyden 2004) in Western society, where we ignore what is 'enough' and continue to demand more and more. Our approach to finite non-renewable resources is thus bizarre and ignores reality.

Our approach to *renewable* resources is even worse. These are in fact the wealth of life present on Earth that gift society so many benefits. Once the world was covered in majestic forests, the plains teemed with grazers and the seas teemed with an abundance of fish. However, while it is possible to sustainably harvest these living 'nature's gifts' (Pascual et al. 2017), overall we have not done so. We have cleared many forests and continue to do so. We overharvest forests beyond sustainable levels. We overfish the oceans and have crashed many fisheries (the classic case being that for the Atlantic Cod, MEA 2005). Meadows et al. (2004: 54) summarise Daly (1990) arguing that:

> For a *renewable resource* (soil, water, forest, fish, etc.), the sustainable rate of use can be no greater than the rate of regeneration of its source.

This is obvious to any ecologist and is the basis of ecologically sustainable forestry and fisheries. Yet in most places this has been ignored or denied (or is a stated policy that is not adhered to). Hence our forests and fisheries continue to shrink, poaching of wild animals expands and mass extinction accelerates. This is clearly not in the long-term interests of nature, *neither* is it in the interests of humanity. The key reason this continues is the endless growth myth (Rees 2008), where if a sustainable harvest of 'X' is good, then a harvest of '10X' would seem to be better. However, such a myth ignores ecological limits and thus leads to ecosystem collapse and species extinction. No extinct species can be harvested 'sustainably'; once gone it is not a renewable resource – just possibly an exhibit in a museum.

The limits to nature's gifts

All the living and non-living things society makes use of, and labels dismissively as 'resources', are in fact the gifts that nature provides for free if used respectfully by society (Washington 2019). However, you do not *coerce* a true gift. If we take a sustainable amount, then the natural world continues to thrive and evolve. If we demand far more, then we push nature beyond what it can gift humanity without degrading and collapsing. We have been demanding too much since the 1980s, when the ecological footprint exceeded what one Earth could provide (GFN 2019). We went into overshoot then (Catton 1982), and despite environmental scientists demonstrating this for decades (see Chapter 1), the response of Western society overall has remarkably been to *demand even more* – and this remains the case today.

Environmental scientists understand that society is fully dependent on nature, but society (in the West at least) has never wanted to accept this (Washington 2013b). Hence the ideas were put forward of 'natural capital' and 'ecosystem services' – to try to show that nature provided so many things to humanity that we were completely reliant on nature. However, these terms (and their definitions) may in fact have caused more problems than they solved. Washington (2019) discusses why this may be so. The key problem is that the definitions of both 'natural capital' and 'ecosystem services' are anthropocentric in nature; they are the services and benefits nature provides *only humanity*. Yet all species require the services their ecosystems provide. Monbiot (2014) argues that natural capital is the triumph of neoliberalism, where we don't speak of 'nature' anymore, for:

> It is now called natural capital. Ecological processes are called ecosystem services because, of course, they exist only to serve us. Hills, forests, rivers: these are terribly out-dated terms. They are now called green infrastructure. Biodiversity and habitats? Not at all à la mode my dear. We now call them asset classes in an ecosystems market.

Space does not permit a more detailed discussion of ecosystem services, but their limitations have been discussed elsewhere (Batavia and Nelson 2017; Washington 2019). Suffice it to say that perhaps it is time to consider that if the worldview and ethics that defined 'ecosystem services' were flawed, the term itself may be also? Of course ecosystem services (or Nature's Contributions to People, Pascual et al. 2017) are not going to go away. However, perhaps we should acknowledge their origins and inherent bias? Perhaps it is time to focus instead on 'People's Contributions to Nature'? Chief among such contributions would be acknowledging nature has intrinsic value, and that society should respect nature and operate from a 'duty of care' towards her.

Growthism and growthmania

William Rees (2010: 5) notes that the: 'flawed assumption that human wellbeing derives from perpetual growth' has distorted the lives of more people than any other cultural narrative in history, and is: 'lodged in the heart of the (un)sustainability conundrum'. Meadows et al. (2004: 12) note that the global economy is already so far above sustainable levels that there is very little time left for the fantasy of an infinite globe. For most of human history, we managed without growth, the last few centuries being the exception (Victor 2008: 24). Enlightened economist John Stuart Mill (1859) thought a 'stationary state' of capital would be a considerable improvement on our present condition. Other enlightened economists such as Schumacher (1973), Daly (1991, 1996, 2014), Victor (2008, 2019) and Jackson (2009) question the desirability and feasibility of continued economic growth (based on increasing population and resource use). The book *Limits to Growth* (Meadows et al. 1972) questioned whether growth could continue forever in a finite world. Lowe (2005) notes it was fiercely attacked because it challenged the fundamental myth of modern society: unlimited growth. Hubbert (1993) explains that during the last two centuries we have known nothing but exponential growth and so have evolved an 'exponential growth culture' dependent on the continuance of exponential growth for its stability. This culture is: 'incapable of reckoning with problems of non-growth' (Daly and Cobb 1994: 408). Daly (in Lowe 2005: 35) says most economists: 'chase an assumption of wants along the road of infinite growth'. Daly (1991: 183) argues that economic growth is unrealistically held to be:

> [t]he cure for poverty, unemployment, debt repayment, inflation, balance of payment deficits, the population explosion, crime, divorce and drug addiction.

Economic growth is thus seen as the *panacea for everything*. Sometimes commitment to growth may be promoted in the guise of 'free trade', 'competitiveness', 'productivity' – or even as 'sustainable development' (Victor 2008: 15). Perhaps nothing stands in the way of healing the world so much as the notion that we can spend or grow our way out of unsustainability (Czech 2013). However, world leaders still seek growth above all else. Indeed, despite all scientific evidence that growthism is the cause of the environmental crisis, politicians continue to advocate for endless growth. Worse, the public continues to elect governments on the basis they will 'keep the economy strong and growing'. Basu (2017) is a former Chief Economist of the World Bank, and argues: 'In 50 years, the world economy is likely to be thriving, with global GDP growing by as much as 20% per year, and income and consumption doubling every four years or so'. Rees (2017) notes about this statement: 'After just 20 years and five doublings, the economy would be larger by a factor of 32; in 50 years it will have multiplied more than 5,000-fold! Basu must inhabit some infinite parallel universe'. The addiction to endless growth thus still remains dominant in society and mainstream economics.

This fixed idea that growth is good is based on false premises and an economic theory that ignores limits and ecological reality. Daly (1991) notes that the verb 'to grow' has become twisted. We have forgotten its original meaning: to spring up and 'develop to maturity' (Ibid.: 99). The original notion included maturity or 'sufficiency', beyond which accumulation gives way to maintenance. Thus growth gives way to maturity, a steady state. Growth is not synonymous with 'betterment'. To grow beyond a certain point can be disastrous. 'Growthmania' is not counting the costs of endless growth (Ibid.).

The idea of endless growth is fundamentally unsustainable. The growth economy is broken and fundamentally unsustainable. Had economists collapsed in deepest shame on being shown in the 1930s (or the 1970s) that their theories fell down against the Second Law of Thermodynamics, we would have made a great deal more progress today towards sustainability (Zencey 2013). However, they did not, and most still in fact refuse to admit that growth economics violates the basic ecological realities of the Earth. Hulme (2009) notes there is a paradox at the heart of economic analyses of climate change, being the presumption of growth. Yet some scholars believe that the fixation on growth and the assumption that 'increasing consumption' is the path to well-being, is *precisely why* we have an environmental crisis. Layard (in Simms et al. 2010: 22) notes:

> Economic growth is indeed triumphant, but to no point. For material prosperity does not make humans happier: the 'triumph of economic growth' is not a triumph of humanity over material wants; rather it is the triumph of material wants over humanity.

The UN Department of Economic and Social Affairs notes that so many of the components of existing economic systems are locked into the use of non-green and non-sustainable technology, while simultaneously so much is at stake in terms of the high cost of moving out of them. The result is 'policy paralysis' (Moore and Rees 2013: 40).

Once we have gone beyond the optimum, and marginal costs exceed marginal benefits, growth will make us worse off. We have then reached *uneconomic growth* (Daly 2008, 2014). However, our experience of diminished well-being will be blamed on 'product scarcity'. The orthodox neoclassical response will then be to advocate increased growth to fix this. In the real world of ecological limits, this will make us even less well off, but will lead to advocacy of even more growth. Daly (1991: 101) paraphrased the original quote of Lewis Carroll (1865) ('The hurrier I go the behinder I get') to: 'The faster we run, the behinder we get' to encapsulate the futility of this process. Daly argues that environment degradation today is largely a disease induced by economic physicians, who treat the sickness of unlimited wants by prescribing unlimited production. But one does not cure a treatment-induced disease by increasing the treatment dosage. Endless growth thus leads to a 'death spiral' (Washington and Kopnina 2018), for as things worsen because of too much growth, the only solution offered by traditional economists and governments is *more* growth. Healing our economy requires accepting the reality that the economy cannot grow forever. As Rees (2017) notes:

> In a non-deluded world, governments would no longer see economic growth as the panacea for all that ails them; in particular, they would acknowledge that enough is literally enough and cease promoting growth as the primary solution to both North-South inequity and chronic poverty within nations.

Regarding growth, Victor (2008: 170) concludes:

> At some point, fundamental questions of growth for what, for whom and with what consequences will be asked by more and more people until there is a shift in societal values away from a growth-first policy. Some glimmers of that shift are discernible today.

I can only reinforce this statement by saying that such glimmers now need to be fanned into a transforming flame.

BOX 3.2 THE FALSE PROMISE OF GREEN GROWTH, PETER A. VICTOR

At a time when the environmental impacts of economic growth are becoming ever more widespread and worrisome, 'green growth' is promoted as the best way forward. Its many proponents contend that green growth offers ever-increasing levels of production and consumption while resource requirements and impacts on the environment are continuously reduced. Green growth has considerable political appeal because it appears to provide a way of avoiding very hard choices between the environment and the economy. Gross domestic product (GDP) can grow forever while resources are conserved, environmental damages decline, ecosystems flourish and all is well.

The concept of 'decoupling' is fundamental to green growth. It means reducing resources and pollutant emissions per unit of GDP fast enough to compensate for the effects of increases in GDP. Just how fast is fast enough depends on the problem at hand. For example, if GDP is to grow at 3% per year for the next 35 years and greenhouse gas (GHG) emissions are to decline by 80%, then GHG emissions per unit of GDP will have to decline on average at over 7% a year. This means that in only 35 years, GHGs emitted per dollar of GDP will have to be more than 90% less than they are now. And if annual GDP growth is to be faster than 3% as many would wish, then the rate of reduction in GHGs per unit of GDP will have to be even faster.

The promise of green growth is further undermined by the fact that environmental pollution and resource depletion are best understood, not as excessive flows, but as the accumulation of unwanted stocks and the decumulation of valuable ones. Unwanted stocks that are accumulating include GHGs in the atmosphere, radioactive wastes from nuclear power plants, and plastics deposited in the oceans. Valuable stocks in decline include rain forests, fisheries, and valuable minerals. This means that even if resource and/or environmental flows are decoupled from growth in GDP, decoupling rates that are too slow will increase unwanted stocks and reduce valuable ones with adverse, unpredictable, consequences.

These considerations are far from theoretical. There is plenty of evidence to show that decoupling at a pace anywhere close to what future economic growth will require, if it is to be green, is quite unprecedented. Even with the best will in the world it is hard to see how the extent and rapidity of decoupling that green growth requires will ever materialize.

A critical choice faces us today, especially those with high incomes. We can cling to the false promise of green growth in the hope that technology will see us through, or we can recognize that fundamental changes are needed in the way societies are organized, how economic output is shared, and in how we relate to others and to the rest of nature. We should reject the false promise of green growth and see how life for all can be improved without relying on economic growth.

> **For further information:**
>
> Victor, P. A. (2019) *Managing without Growth: Slower by Design, not Disaster,*
> Cheltenham, UK: Edward Elgar Publishing.
> Also see: www.pvictor.com

Conclusion

Accepting reality should be the first thing we do – for we live in the *real* world.
To ignore it is to ignore the risks we run and possibly to put survival on the line.
Many past civilizations did just this and hence collapsed (Diamond 2005). Do we
want to follow suit? This is certainly a possibility to consider (see Chapter 10),
but I would suggest any thinking and compassionate person would like to avoid
this. That means we need to stop denying our dire predicament. Accepting reality
means we have to abandon hubris and the idea that the world is all about us.
Even more, it means breaking the denial dam about our problems – but also the
solutions to them. For example, many who deny climate change simultaneously
deny renewable energy as a key solution (Washington 2015). Finally, accepting
reality means denying the possibility of endless physical growth on a finite
planet – especially one where environmental crisis and ecocide are being driven
by that endless growth fantasy. It means ceasing to pretend that resources will
never be limiting and seeking to conserve our minerals (commonly called non-
renewable resources) as well as the living world (commonly called renewable
resources). There is thus a lot for us to accept and a lot of denial to roll back.
However, once we accept our problems we can start to powerfully act on the
solutions to those problems. Acceptance of our problems would be made so
much easier if we changed our worldview from anthropocentrism to ecocentrism
(see Chapter 2). If one upholds an ecocentric worldview, one could only ethically
seek to heal ecocide, to act to create an ecologically sustainable future.

Yet if a large part of the public abandons denial, they can fairly quickly turn
around corporate denial, especially if it reduces the profits of corporations. If
a large part of the community tells our politicians that they want real action
now (not weasel words), then politicians will actually *act*. We are not powerless
drones who cannot change things. En masse, if we accept the 'Great Work'
(Berry 1999) of repairing the Earth, we have the vision, the creativity and the
power to solve the environmental crisis and to heal our world (Washington
2015). That nobody should deny.

What can *I* do?

1)   Accept and talk about the limits clearly shown by environmental science
     and summarised in Chapter 1. If someone you know suggests there are 'no
     limits', speak up politely to explain why they are mistaken.

2) Recognise that all of us deny something, and think about your own denial in regard to things that wound the world. The cure for denial is *dialogue*. Thus don't shy away from talking (politely) about the need to heal the world. Web resources are available to help you carry out such polite dialogues, such as CRP (n.d.) and PSC (n.d.). In such dialogue it is important to stay positive (CC 2018). After all, there *are* positive solutions; they may be challenging but they are completely feasible (see Chapters 10 and 11).

3) Consider *your own* impact. For example, do you commonly fly overseas for work or holidays? This is common in society, but it causes a huge carbon footprint that accelerates climate change. If you do have to fly (though often one doesn't), do you always buy reliable 'gold standard' carbon offsets (e.g. https://www.atmosfair.de/en/)?

4) Have you assessed *your own* ecological footprint (http://www.footprintcalculator.org/) and considered actions to improve this (such as eating less meat and flying less)?

5) Argue for a truly 'honest market' that factors in all environmental and social costs. Reuse and recycle products. If possible grow your own food. Argue against rampant consumerism; concentrate on well-being rather than more 'stuff'. Consume less, shop less and live more!

6) Apply the phrase 'Reduce, Reuse, Recycle' but add a 'Rethink' to this. Rethink whether you really *need* a new car or other product.

7) Speak out against endless growth and growthmania. Point out that on a finite planet endless physical growth is irrational and unsustainable, as it has become our major destructive addiction. Whenever you hear someone saying 'Our economy *must* grow!' don't just nod, point out that nothing in nature grows forever, and we need a steady state economy, not a destructive endless growth economy.

References

Abbey, E. (1977) *The Journey Home*, New York: Dutton.

Almeida, P. and Silva, P. (2009) 'The peak of oil production – Timings and market recognition', *Energy Policy*, **37** (4): 1267–1276.

Arndt, N. T., Fontbote, L., Hedenquist, J. et al. (2017) 'Future global mineral resources', *Geochemical Perspectives*, **6** (1): 1–171.

Assadourian, E. (2013) 'Re-engineering cultures to create a sustainable civilization', in *State of the World 2013: Is Sustainability Still Possible?* ed. L. Starke, Washington: Island Press, pp. 113–125.

Barry, J. (1994) 'The limits of the shallow and the deep: Green politics, philosophy and praxis', *Environmental Politics*, **3**: 369–394.

Basu, K. (2017) 'The global economy in 2067', Project syndicate website Jun 21st, see: www.project-syndicate.org/commentary/long-term-global-economic-prospects-by kaushik-basu-2017-06?barrier=accesspaylog (accessed 2 May 2019).

Batavia, C. and Nelson, M. P. (2017) 'For goodness sake! What is intrinsic value and why should we care?', *Biological Conservation*, **209**: 366–376.

Berry, T. (1999) *The Great Work: Our Way into the Future*, New York: Bell Tower.

Boyden, S. (2004) *The Biology of Civilisation: Understanding Human Culture as a Force in Nature*, Sydney: UNSW Press.

Braungart, M. and McDonough, W. (2008) *Cradle to Cradle: Remaking the Way We Make Things*, London: Vintage Books.

Brown, L. (1978) *The 29th Day*, Washington: Norton and Co.

Brown, L. (2011) *World on the Edge: How to Prevent Environmental and Economic Collapse*, New York: W.W. Norton and Co.

Bryant, P. (1995) 'Constructing nature again', in *Main Currents in Western Environmental Thought*, ed. P. Hay, Sydney: UNSW Press, pp. 24–25.

Burr, V. (2003) *Social Constructionism*, Second edition, New York: Routledge.

Butler, C. (2002) *Postmodernism: A Very Short Introduction*, Oxford/New York: Oxford University Press.

Cardinale, B. J., Duffy, E., Gonzalez, A. et al (2012) 'Biodiversity loss and its impact on humanity', *Nature*, **486** (7401): 59–67, DOI: 10.1038/nature11148

Carroll, L. (1865) *Alice in Wonderland*, London: Macmillan.

Catton, W. (1982) *Overshoot: The Ecological Basis of Revolutionary Change*, Chicago: University of Illinois Press.

CC. (2018) '5 reasons to stay positive in the face of climate inaction', Climate Council (Australia), see: www.climatecouncil.org.au/5-reasons-to-stay-positive-in-the-face-of-climate-inaction/?gclid=Cj0KCQiApbzhBRDKARIsAIvZue9WWsjAz00ySrfBPic-pCbP088t7L95IHPqFEo2Kkyvb3g3wWtPZAgaAm3cEALw_wcB (accessed 27 March 2019).

Cooper, A., Brown, T., Price, S. et al. (2018) 'Humans are the most significant global geomorphological driving force of the 21st century', *The Anthropocene Review*, **5** (3): 222–229.

Crist, E. (2012) 'Abundant earth and the population question', in *Life on the Brink: Environmentalists Confront Overpopulation*, eds. P. Cafaro and E. Crist, Georgia: University of Georgia Press, pp. 141–151.

CRP (n.d.) 'How to talk to your friends about the climate crisis – According to science', The Climate Reality Project, see: www.climaterealityproject.org/blog/how-talk-your-friends-about-climate-crisis-according-science (accessed 27 March 2019).

Curry, P. (2011) *Ecological Ethics: An Introduction*, Second edition, Cambridge: Polity Press.

Czech, B. (2013) *Supply Shock: Economic Growth at the Crossroads and the Steady State Solution*, Canada: New Society Publishers.

Daly, H. (1990) 'Toward some operational principles of sustainable development', *Ecological Economics*, **2**: 1–6.

Daly, H. (1991) *Steady State Economics*, Washington: Island Press.

Daly, H. (1996) *Beyond Growth: The Economics of Sustainable Development*, Boston: Beacon Press.

Daly, H. (2008) 'A steady-state economy: A failed growth economy and a steady-state economy are not the same thing; they are the very different alternatives we face', 'thinkpiece' for the Sustainable Development Commission, UK, April, 24, 2008, see: http://steadystaterevolution.org/files/pdf/Daly_UK_Paper.pdf (accessed 27 March 2019).

Daly, H. (2014) *From Uneconomic Growth to the Steady State Economy*, Cheltenham: Edward Elgar.

Daly, H. and Cobb, J. (1994) *For the Common Good: Redirecting the Economy Toward Community, the Environment, and a Sustainable Future*, Boston: Beacon Press.

Diamond, J. (2005) *Collapse: Why Societies Choose to Fail or Succeed*, New York: Viking Press.

Ehrlich, P. and Ehrlich, A. (1991) *Healing the Planet: Strategies for Resolving the Environmental Crisis*, New York: Addison–Wesley Publishing Company.

Ehrlich, P. and Ornstein, R. (2010) *Humanity on a Tightrope: Thoughts on Empathy, Family and Big Changes for a Viable Future*, New York: Rowman and Littlefield.

Eliot, T. (1936) *Collected Poems, 1909–1935*, London: Faber and Faber.

EM. (2015) 'The Ecomodernist Manifesto', see: www.ecomodernism.org/manifesto-english/ (accessed 27 March 2019).

Gare, A. (1995) *Postmodernism and the Environmental Crisis*, London/New York: Routledge.

GFN. (2019) 'World footprint', Global Footprint Network, see: www.footprintnetwork.org/en/index.php/GFN/page/world_footprint/ (accessed 27 March 2019).

Giddens, A. (1994) *Beyond Left and Right: The Future of Radical Politics*, Palo Alto, CA: Stanford University Press.

Giddens, A. (2009) *Global Politics and Climate Change*, Cambridge: Polity Press.

Gilder, G. (1981) *Wealth and Poverty*, New York: Bantam.

Hansen, J. (2009) *Storms of My Grandchildren: The Truth about the Coming Climate Catastrophe and Our Last Chance to Save Humanity*, London: Bloomsbury.

Harich, J. (2010) 'Change resistance as the crux of the environmental sustainability problem', *System Dynamics Review*, **26** (1): 35–72.

Heidegger, M. (1955) *Martin Heidegger: Philosophical and Political Writings*, New York: Continuum International.

Heron, J. (1992) *Feeling and Personhood: Psychology in Another Key*, London: Sage.

Hubbert, K. (1993) 'Exponential growth as a transient phenomenon in human history', in *Valuing the Earth: Economics, Ecology, Ethics*, eds. H. Daly and K. Townsend, Cambridge, MA: MIT Press, pp. 113–126.

Hulme, M. (2009) *Why We Disagree About Climate Change: Understanding Controversy, Inaction and Opportunity*, Cambridge: Cambridge University Press.

IPCC. (2018) *Global Warming of 1.5 C*, International Panel on Climate Change, see: www.ipcc.ch/pdf/special-reports/sr15/sr15_spm_final.pdf (accessed 27 March 2019).

Jackson, T. (2009) *Prosperity without Growth: Economics for a Finite Planet*, London: Earthscan.

Laing, R. D. (1971) *The Politics of the Family*, New York: Penguin.

Lifton, R. and Falk, R. (1982) *Indefensible Weapons*, New York: Basic Books.

Lopez, B. (1986) *Arctic Dreams*, New York: Scribners.

Louv, R. (2005) *Last Child in the Woods: Saving our Children from Nature-Deficit Disorder*, London: Atlantic Books.

Louv, R. (2011) *The Nature Principle: Reconnecting with Life in a Virtual Age*, North Carolina: Algonquin Books.

Lowe, I. (2005) *A Big Fix: Radical Solutions for Australia's Environmental Crisis*, Melbourne: Black Inc.

Malthus, T. R. (1798) *An Essay on the Principle of Population*, London: J. Johnson.

McCright, A. and Dunlap, R. (2000) 'Challenging global warming as a social problem: An analysis of the conservative movement's counter-claims', *Social Problems*, **47** (4): 499–522.

MEA. (2005) *Living Beyond Our Means: Natural Assets and Human Wellbeing, Statement from the Board, Millennium Ecosystem Assessment*, United Nations Environment Programme, see: www.millenniumassessment.org.

Meadows, D., Meadows, D., Randers, J. and Behrens, W. (1972) *The Limits to Growth*, Washington: Universe Books.

Meadows, D., Randers, J. and Meadows, D. (2004) *Limits to Growth: The 30-Year Update*, Vermont: Chelsea Green.

Michaux, S. (2014) 'The coming radical change in mining practice', in *Sustainable Futures*, eds. J. Goldie and K. Betts, Canberra: CSIRO Publishing, pp. 73–84.

Mill, J. S. (1859) *On Liberty*, Chicago: Encyclopedia Britannica Great Books (1952, originally published 1859).

Monbiot, G. (2014) 'Put a price on nature? We must stop this neoliberal road to ruin', *The Guardian* 24 July 2014, see: www.theguardian.com/environment/georgemonbiot/2014/jul/24/price-nature-neoliberal-capital-road-ruin (accessed 8 July 2018).

Moore, J. and Rees, W. (2013) 'Getting to one-planet living', in *State of the World 2013: Is Sustainability Still Possible?*, ed. L. Starke, Washington: Island Press, pp. 39–50.

Nordhaus, W. and Tobin, J. (1973) 'Is growth obsolete?', in *The measurement of Economic and Social Performance*, ed. M. Moss, New York: National Bureau Economic Research, pp. 509–564.

Oreskes, N. and Conway, M. (2010) *Merchants of Doubt: How a Handful of Scientists Obscured the Truth on Issues from Tobacco Smoke to Global Warming*, New York: Bloomsbury Press.

Pascual, U., Balvanera, P., Diaz, S. et al. (2017) 'Valuing nature's contributions to people: The IPBES approach', *Current Opinion in Environmental Sustainability*, **26–27**: 7–16.

Plumwood, V. (2003) 'New nature or no nature?', Unpublished paper provided by the author, 14/8/03. This paper has since been incorporated in Plumwood, V. (2006 forthcoming), The Concept of a Cultural Landscape, *in* 'Ethics and the Environment' (forthcoming 2006), Indiana University Press, Bloomington, IN.

PSC. (n.d.) 'Psychology for a safe climate', see: www.psychologyforasafeclimate.org/untitled (accessed 17 March 2019).

Reason, P. and Torbert, W. R. (2001) 'Toward a transformational social science: A further look at the scientific merits of action research', *Concepts and Transformations*, **6** (1): 1–37.

Rees, W. (2008) 'Toward sustainability with justice: Are human nature and history on side?', in *Sustaining Life on Earth: Environmental and Human Health through Global Governance*, ed. C. Soskolne, New York: Lexington Books, pp. 81–94.

Rees, W. (2010) 'What's blocking sustainability? Human nature, cognition and denial', *Sustainability: Science, Practice and Policy*, **6** (2) (ejournal), see: http://sspp.proquest.com/archives/vol6iss2/1001-012.rees.html (accessed 21 March 2019)

Rees, W. (2017) 'Staving off the coming global collapse', *The Tyee* 17th July, see: https://thetyee.ca/Opinion/2017/07/17/Coming-Global-Collapse/ (accessed 1 May 2019).

Rees, W. (2019) 'End game: The economy as eco-catastrophe and what needs to change', *Real World Economic Review*, **87**: 132–148.

Rolston, H., III. (2012) *A New Environmental Ethics: The Next Millennium for Life on Earth*, New York: Routledge.

Schumacher, E. (1973) *Small Is Beautiful*, New York: Harper Torchbooks.

Sharot, T. (2011) 'The optimism bias', *Current Biology*, **21** (23): R941–R945.

Simms, A., Johnson, V. and Chowla, P. (2010) *Growth Isn't Possible: Why We Need a New Economic Direction*, London: New Economics Foundation, see: www.neweconomics.org/publications/growth-isnt-possible (accessed 8 March 2018).

Simon, J. (1981) *The Ultimate Resource*, New Jersey: Princeton University Press.

Soskolne, C. (2008, ed.) *Sustaining Life on Earth: Environmental and Human Health through Global Governance*, New York: Lexington Books.

Soule, M. (1995) 'The social siege of nature', in *Reinventing Nature: Responses to Postmodern Deconstruction*, eds. M. Soule and G. Lease, Washington, DC: Island Press, pp. 137–170.

Stiglitz, J. (1979) 'A neoclassical analysis of the economics of natural resources', in *Scarcity and Growth Reconsidered*, ed. K. Smith, Baltimore: Johns Hopkins University Press, pp. 36–66.

Sukhdev, P. (2010) 'Preface', in *The Economics of Ecosystems and Biodiversity: Ecological and Economic Foundations*, ed. P. Kumar, London: Earthscan, pp. xvii–xxviii.

Suzuki, D. (2010) *The Legacy: An Elder's Vision for Our Sustainable Future*, Vancouver: Greystone Books.

UNEP. (2011) *Recycling Rates of Metals: A Status Report*, Nairobi: United Nations Environment Programme.

Vetlesen, A. (2015) *The Denial of Nature: Environmental Philosophy in the Era of Global Capitalism*, London: Routledge.

Victor, P. (2008) *Managing Without Growth: Slower by Design, not Disaster*, Cheltenham, UK: Edward Elgar.

Victor, P. (2019) *Managing without Growth: Slower by Design, not Disaster*, Second edition, Cheltenham, UK: Edward Elgar.

von Weizsäcker, E., Hargroves, K., Smith, M., Desha, C. and Stasinopoulos, P. (2009) *Factor 5: Transforming the Global Economy through 80% Increase in Resource Productivity*, UK: Earthscan.

Washington, H. (2013a) *Poems from the Centre of the World*, lulu.com, see: www.lulu.com/shop/haydn-washington/poems-from-the-centre-of-the-world/paperback/product-21255751.html (accessed 27 March 2019).

Washington, H. (2013b) *Human Dependence on Nature: How to Help Solve the Environmental Crisis*, London: Earthscan.

Washington, H. (2015) *Demystifying Sustainability: Towards Real Solutions*, London: Routledge.

Washington, H. (2018a) 'Denial – The key barrier to solving climate change', in *Encyclopedia of the Anthropocene*, eds. D.A. DellaSala and M. I. Goldstein, UK: Elsevier, pp. 493–499.

Washington, H. (2018b) *A Sense of Wonder Towards Nature: Healing the Planet through Belonging*, London: Routledge.

Washington, H. (2019) 'Ecosystem services – A key step forward *or* anthropocentrism's "Trojan Horse" in conservation?', in *Conservation: Integrating Social and Ecological Justice*, eds. H. Kopnina and H. Washington, New York: Springer, pp. 73–90.

Washington, H., Chapron, G., Kopnina, H., Curry, P., Gray, J. and Piccolo, J. (2018) 'Foregrounding ecojustice in conservation', *Biological Conservation*, **228**: 367–374.

Washington, H. and Cook, J. (2011) *Climate Change Denial: Heads in the Sand*, London: Earthscan.

Washington, H. and Kopnina, H. (2018) 'The insanity of endless growth', *The Ecological Citizen*, **2** (1): 57–63.

Zencey, E. (2013) 'Energy as a master resource', in *State of the World 2013: Is Sustainability Still Possible?*, ed. L. Starke, Washington: Island Press, pp. 73–83.

Zerubavel, E. (2006) *The Elephant in the Room: Silence and Denial in Everyday Life*, London: Oxford University Press.

4

THE TRAGEDY OF OVERPOPULATION DENIAL

To live in
Balance with life
Demands we know when
Enough is enough.
It implies reciprocity,
Requires respect
For the life
That sustains us.
It means each takes
Only so much.
Balance means
A sustainable number
Of we humans
Live on land
In harmony …

From 'Enough' (Washington 2019)

SUMMARY

One of the key problems behind why the world is wounded is that it is *over-populated* by humanity. There are too many people using too many resources, with the wrong technology (fossil fuels). We need to reduce the size of our ecological footprints, but we also need *less feet*. Continuing population growth exacerbates all environmental problems. Global population is at 7.7 billion people but is projected to reach 9.8 billion by 2050 and 11.2 by 2100. Yet estimates of what an ecologically sustainable global population might be range from 1 to 4 billion. Environmental scientists use the equation Environmental Impact = Population × Affluence × Technology, and argue we must act on *all* these issues. A key problem is that many in society ignore or deny the problem of overpopulation (for several reasons that are discussed). Hence the environmental crisis worsens, which is bad for both nature and humanity (and is leading us towards what can only be called a 'tragedy in the making'). Ignoring overpopulation means we actually ignore the well-being of future generations (human and nonhuman). However, several humane (non-coercive) solutions exist to stabilise (and then reduce) population. We just need to stop denying the issue. For the detail explaining this summary, please read on; however, if you just want to know the things *you can do* (eight are listed) about this, then turn to the end of the chapter.

Introduction

While overpopulation is part of the issue of ecological limits (Chapters 1 and 3), it is so important that it requires a chapter of its own. Why? Because it is so often ignored or denied by both academia and mainstream society. Indeed, even otherwise 'green' and environmentally thoughtful people continue to ignore this key issue. The key problem is that the world is overpopulated. This is true globally, as well as for any country which has exceeded ecological limits (see Chapter 1). Hence the world *cannot* sustainably support further growth in population beyond replacement levels (i.e. beyond two children per couple).

Population growth exacerbates all environmental problems and is a key underlying aspect of the endless growth economy (Chapter 3). Apart from daring to question the growth economy, nothing else seems to raise such outrage in mainstream society as suggesting we should 'limit human numbers'. Into it comes issues such as religion, racism, social and ecological justice, equity and poverty (Cafaro and Crist 2012; Kopnina and Washington 2016). There is also no more 'taboo' issue politically than population. Collectively, the public and governments have been shying away from it for decades. Yet Hulme (2009) notes that if there is a 'safe' level of greenhouse gases to avoid runaway climate change, then is there not also a desirable world population? We shall examine the predicament of human overpopulation and why society so often ignores or denies it.

It is not just footprints: it's too many feet

Unsustainable population growth pushes the world beyond its carrying capacity (Catton 1982), being the number of people an area can support sustainably and indefinitely. The world is finite, and we know that human numbers have grown exponentially so that they are now far larger than ever before in history. Our global population is more than 7.7 billion people. Despite the declining global

Total Fertility Rates (TFRs), population momentum is projected to cause global population to rise to 9.8 billion by 2050 and 11.2 billion by 2100 (UNDESA 2017). When I was born in 1955 world population was *2.8 billion*, a level that was just about ecologically sustainable. Today in 2019 it is *7.7 billion* (see Figure 4.1), having increased by a factor of 2.7 times – just in my lifetime. As the environmental crisis discussed in Chapter 1 has shown, it is now way beyond an ecologically sustainable level.

So the idea commonly trotted out that the population explosion is 'over' (often stated by the media and some scholars) is clearly wrong (as discussed by Campbell 2012). However, it is also worth pointing out that some biologists challenge the UNDESA (2017) projections, arguing that future population may well be *larger* than that suggested (Collins and Page 2019). Chapter 1 provides the data that shows that the world is in ecological overshoot, with massive extinction underway – as a result of our current 7.7 billion people. Adding another 2.1 billion by 2050, and 3.5 by 2100, would cause further massive impact, major clearing of native vegetation (to produce food), major escalation of greenhouse gas production, major ecosystem collapse and an even greater mass extinction (Crist et al. 2017). The full ramifications of overpopulation are well shown in the 'Human Overpopulation Atlas' by Joao Abegao (2018).

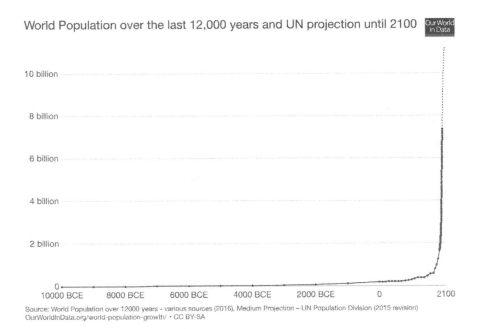

Source: World Population over 12000 years - various sources (2016), Medium Projection – UN Population Division (2015 revision) OurWorldInData.org/world-population-growth/ • CC BY-SA

FIGURE 4.1 World Population growth over time

(Source: https://ourworldindata.org/world-population-growth)

In 1968 Paul Ehrlich published *The Population Bomb*, which alerted the world to the dangers of exponentially growing population. He was later part of coining the entity (Ehrlich et al. 1977):

Environmental **I**mpact = **P**opulation × **A**ffluence × **T**echnology

Or I = PAT. Our impact on the Earth is thus the number of people times their affluence (per capita consumption of resources) times the technology we use. Of course, like most environmental scientists, I accept that historically most of the impact from pollution and carbon emissions has come from the consumers in the developed world (Monbiot 2009). However, the developing world is rapidly catching up. If this is done using traditional carbon-polluting industry (as it mostly still is), then the result will be steeply accelerating global carbon emissions, resource consumption and increasing pollution. Indeed, this is already happening. The technology used to 'catch up' will thus be a critical factor (see Chapter 6), as will the question of whether the developing world seeks to catch up to the incredibly wasteful US (or Australian) level. However, improving technology or reducing affluence can only reduce our impact so far. In the end the numbers of people themselves count. The sheer number of consumers matters as much as the fact that many are now consuming more (Washington 2015). A big population has a big impact, especially as the developing world expands its economy. Despite a 30% increase in resource efficiency, global resource use has expanded by 50% over 30 years (Flavin 2010). This is mainly due to the increasing affluence of the large populations in the developing world. This is why China is now the world's biggest carbon polluter, while India now ranks third (after the United States at number two) (UCS 2018). Accordingly, we need to target *all three components* of I = PAT if we seek to reduce human impact: reducing population, limiting affluence and cleaning technology. These are key tasks of any meaningful interpretation of what 'healing the planet' should mean if we are to reach a sustainable future. Ignoring any one part of I = PAT (such as population) will not work if we truly want to heal the world.

What is an ecologically sustainable global population?

Washington (2015) summarises the views on what an ecologically sustainable population number for the Earth might be. Crist (2012) points out that this question should really be what is the number (and at what level of consumption) that can live on Earth without turning it into a human colony founded on the genocide of the nonhuman. Biocapacity data suggest that if we made no change at all to consumption patterns, we could currently sustain a population of 4 to 5 billion. Our ecological footprint suggests no more than 4.7 billion people (Engelman 2013), but not if every one of those lived at the US standard, where the Earth could sustain only a quarter of today's population, or 1.75 billion people (Assadourian 2013). If we were to move to the European standard of consumption it has been argued it would be 2 billion (WPB n.d.). If everybody on Earth shared a modest standard of living, midway between the richest and the poorest, that figure might be around 3 billion (PM 2010). The world is clearly already *over*populated in regard to sustainability. We cannot live in harmony with nature when our numbers are degrading the world's life support systems and causing ecocide – ecological genocide (Washington 2015). It is also worth considering that human actions have already degraded the ability of the Earth to support people. This is made clear by the Millennium Ecosystem Assessment (MEA 2005), which stated that 60% of ecosystem services were degrading or being used unsustainably. Hence Prof Paul Ehrlich (2013) comments that he used to think the ecologically sustainable world population was 2 billion, but given the ecocide society has caused, he now thinks it is more likely 1 billion. More recently he has been quoted as saying it is 1.5 billion (Carrington 2018).

To feed all the world's people by 2050 (given rising populations and incomes) food production must increase by 70% according to a FAO report, but at the same time the report noted 25% of the world's land is degraded, and water is becoming increasingly scarce and polluted (both above and below ground) (FAO 2011). It is hard to see how it is possible in the future to adequately feed 9 or 11 billion people, given the many accelerating and interconnected environmental problems that food production now faces (Brown 2012). It might perhaps be possible by destroying the majority of our remaining natural areas and biodiversity to increase cropland by a fifth (Erb et al. 2009). This would cause massive negative impacts on nature (Crist et al. 2017). Engelman (2012) argued in 2012 we could stabilise world population at 8 billion if we applied the strategies suggested later. Society didn't act then, so Engelman (2016) suggests we could halt population growth before mid-century at a level below 9 billion. Then we would need to reduce overall population over time towards a lower, ecologically sustainable number, at least half that (possibly 1–2 billion would be a more responsible and ecologically realistic figure, Washington 2015). While population remains in the 7–8 billion range, greenhouse gas production and material use will have to come way down from Western levels if we are to rapidly reduce the overall impact (Engelman 2013).

BOX 4.1 HOW TO RAISE A CHILD TO HEAL A WOUNDED WORLD, ERIK ASSADOURIAN

The choice to have children on a rapidly changing planet is not one to make lightly. But for many people, one of the great joys of life is having a child, regardless of how bad things are going to get. As a sustainability researcher, lead author of *Earth Education: Rethinking Education on a Changing Planet*, and father of one 7-year-old son, I have thought long about what one should teach children to be the leaders ready to guide the sustainability transition *and* resilient enough to adapt and survive as the planetary crisis unfolds.

First, having a child is the choice that has the single-most ecological impact of anything you do in your life. Ideally, keeping in mind that we need to reduce human population, and instead we're heading to 9.6 billion, you'll consider having only one. That will help solve the sustainability crisis more than nearly any other action you can take (and you can always adopt a second if you want a larger family). Once you have a child, raise her consciously, especially in the first two formative years of life. Feed her healthy, sustainable food. My wife and I kept our son, Ayhan, away from sweets and screens for his first two years. And since he never had them, he never knew what he was missing. For our own sanity that was worth it (two years free from nagging!), but it's also great for children's physical and mental development, and sets a path for healthier relationships with both of these treats.

At all ages, get your child outside and into nature – this could be a backyard, garden or pocket park if a nature reserve isn't reachable. My son started helping me garden and compost (in our four square meter urban yard) once he started walking, and we continue to garden together today. Ayhan also now spends his Tuesdays in the forest, with a group of other kids and nature mentors, learning ancestral skills and forming a deeper connection to Gaia than I had at that age. Forming that relationship is as essential as having a cerebral understanding of science behind the planetary crisis (which we work on through many hours of nature documentaries, readings, and stories).

Beyond ecological connection and understanding, it is essential that we raise our children to be strong moral leaders. Some countries, like Singapore, integrate moral education directly into school, but most countries don't. Talk about morality, doing good, manners, and responsibilities from an early age. Considering the incredibly difficult moral decisions all of us will be forced to make as things fall apart, building this foundation at a young age is paramount. As well, teach your child essential life skills – languages, how to cook, how to grow food, first aid, martial arts. All of these bring joy and opportunity today, and in the future, as societies simplify and consumer luxuries become scarce – will become even more valuable. Plus, while savings often evaporate in societal collapse, skills remain.

More than anything else, teach your child to be a leader in service to Earth. Use the knowledge, skills, and social emotional learning they have acquired to plant, protest, heal, rebuild, and guide humanity to a more sustainable, healthier, and more secure future. Invite them to find their way to do that from an early age. And reinforce, at all ages, that this service, more than anything else, is what makes a life worth living.

For further information: Worldwatch (2017) *EarthEd: Rethinking Education on a Changing Planet*, Washington: Island Press.

Why is overpopulation such a difficult policy issue?

Why is population such a diabolical policy issue? Because it cuts at the heart of the received wisdom of a million years of human evolution, where 'more' people was always seen as being better (Washington 1991). 'More' meant we could gather more food, cut down more forest to grow food, hunt more animals, defend ourselves better. 'More people' as a concept (until the last 100 years) has always been seen as a 'good thing' for society (Washington 2015). Clearly, people love babies, so it goes against the grain to say we should have fewer of them. Even authors of sustainability classics such as *Cradle to Cradle* (Braungart and McDonough 2008) balk at talking about stabilising population. However, Collins (2010) believes that at the core of the population problem is a 'conflict of rights' – the right of the individual to reproduce, and the right of other species to continue to exist. It is very hard for us to understand in our hearts that now 'more' is no longer better. Ehrlich (2017) notes:

> The more people there are, the more products of nature they demand to meet their needs and wants: timber, seafood, meat, gas, oil, metal ores, rare earths and rare animals to eat or to use for medicinal purposes. Human demands cause both habitat destruction and outright extermination of wildlife. So when you watch the expansion of the human enterprise; when you see buildings springing up; when you settle down to dinner at home or in a restaurant; you are observing (and often participating in) the sixth mass extinction.

Add to this the religious discouragement of birth control methods (e.g. the Catholic Church). Add to that the fundamental desire of governments to have more citizens and greater power. Population ecologist Meyerson (see Hartmann et al. 2008) explains:

> Conservatives are often against sex education, contraception and abortion and they like growth – both in population and in the economy. Liberals usually support individual human rights above all else and fear the coercion label and therefore avoid discussion of population growth and stabilisation. The combination is a tragic stalemate that leads to more population growth.

History shows a worrying decline in discussion of overpopulation by the UN and others (Campbell 2012). In 1994 the UN 'Cairo' conference stopped talking about 'family planning' and instead spoke only of 'women's reproductive health' (funding for family planning then dropped worldwide). At that time population control became something of a taboo word, as it was portrayed as infringing on 'women's rights'. Many in the Left have referred to the past short-lived and unsuccessful forced sterilisation programme in India, suggesting (erroneously) that most family planning was coercive (Ibid.). In fact family planning is about *giving women the choice* as to when to use their 'right' to have children. In fact, if family planning and contraceptives were made universally available, the evidence is that population growth would stabilise and then start to decline (Engelman 2012). Another problem has been a common (if not universal) trend in feminism and the political Left to argue against population control, claiming it is coercive (Beck and Kolankiewicz 2001), though this may be starting to change (Weeden and Palomba 2012). However, Weld (2012) speaks of:

> The Marxist/feminist/social justice ideology that denies the population factor and vilifies those who address it … An ideology that claims to promote social justice, but does not recognize that the Earth is finite, is more than unethical.

Population exacerbates all other environmental problems (Washington 2015). Butler (2012) notes that both climate change and the extinction crisis are merely *symptoms* of ecological overshoot by an obese humanity. Rees (2019: 132) argues that all our environmental problems are: 'symptoms of a singular phenomenon, *gross human ecological dysfunction*'. The IPCC 'Climate Change 2014' Synthesis report (2014: 5) itself has noted:

> Globally, economic and population growth continued to be the most important drivers of increases in CO_2 emissions from fossil fuel combustion.

The IPBES (2019) extinction report similarly notes that population growth is a key cause of species extinction. Interestingly, I have found that even some climate scientists and activists fail to understand that climate change is actually a symptom of overpopulation and overconsumption. Hence, some climate scientists and activists still will not tackle either economic or population growth. Overpopulation also means cutting more forest for farmland, over-farming land so that it erodes, killing more 'bush meat' (wild animals) for food, and over-fishing the rivers and seas (Washington 2015). It means burning more fossil fuels, or clearing and burning more forest as a way of fuelling 'development'. Many scholars write of the need for a 'smaller ecological footprint', but as Dietz and O'Neill (2013: 78) point out: 'we need smaller footprints, but we also need fewer feet'. And yet despite all this, talking about overpopulation remains controversial. As Chris Rapley, the Director of London's Science Museum has noted:

> So controversial is the subject that it has become the 'Cinderella' of the great sustainability debate – rarely visible in public, or even in private.
>
> *(BBC 2006)*

Other reasons why we deny overpopulation

Kopnina and Washington (2016) note that in practice, many governments actually try to boost their population growth by pro-child policies in order to boost their economic and political advantage over their less populous neighbours. In some countries, population growth is seen favourably, as politicians and economists assume that a larger population stimulates economic growth, both in terms of markets and consumers (*The Economist* 2012a, 2012b; Blowfield 2013). In the 2010s, population growth has become a polarised issue in sustainability discourse (Kopnina and Washington 2016), with debates ranging from ambiguity to open hostility towards 'blaming' overpopulation. Polarisation into the 'guilty' high-consumption Westerners and 'poor victims' in developing countries (who are portrayed as not being able to help having many children, and who have a much smaller carbon footprint due to their poverty) leads several scholars to claim population action is 'racist' or socially unfair (e.g. White 1994; Hartmann 2004; Robbins 2012; Fletcher et al. 2014). Illustrating this anti-population control sentiment, White (1994) blames green politics for what he sees as vilifying population growth.

During the 1972 UN Conference on the Human Environment, international agendas integrating population and sustainability were proposed. These international agendas challenged the fundamental fallacy infecting industrial capitalism, that unlimited growth both of population and the economy is possible on a planet of finite resources. However, recently, the linking of population and sustainability has become controversial (Kopnina and Washington 2016). Policy documents issued by the UN's 'Sustainable Development Goals' (SDGs) do not seriously address population issues (Ibid.). At the UN Conference on Sustainable Development (also called Rio+20) in 2012, the key problem discussed was a concern with agricultural productivity and efficiency, and the necessity to provide food for a growing population (but not discussing any focus on stabilizing and reducing population). More recently, the idea of 'planetary boundaries' (Rockstrom et al. 2009; Steffen et al. 2015) sadly failed to consider 'population' as a planetary boundary, even though it directly impacts negatively on *all* such boundaries. It is worth noting also that a book on a new ecological economic model, *Doughnut Economics* (Raworth 2017), mentions population a few times, but then dismisses it as an issue, stating that population growth rates are declining (which is largely true outside Africa). It fails to note that population itself is *still on the rise* due to population momentum, however, an ongoing reason why the world has overshot four planetary boundaries. Even the recent Living Planet Index (WWF 2018) has marginalised population as a key issue, instead focusing on consumption. Also, some 'degrowth' advocates paradoxically (to my mind) are arguing that overpopulation is not a key issue. Some argue that net immigration should not be considered in terms of its contribution to environmental impact, and argue for an 'open door policy' (Kallis 2018). Derer (2018) notes: 'Nowadays there is almost a complete silence about overpopulation, both in the media and academia'.

Kenneth Smail (2003, 2016) reflects that despite the impossibility of decoupling population from sustainability concerns, the issue of population growth has gained a certain 'political charge'. Population growth is shunned in 'politically correct' academic circles, with critics arguing that we do not have a global overpopulation issue, but a global issue of overpopulation of just the highly privileged and exploitative Western minority, and that population growth is used as a scapegoat by rich over-consuming elites (Fletcher 2014). In this critique, those who link population to sustainability are branded neo-Malthusian, racist or misanthropic. While the 'Northern' populations are only growing slowly (and consumption is fairly static), the 'Southern' countries, particularly in sub-Saharan Africa, are still growing rapidly in population. The ethical question posed is that the global 'North' or 'West' should not tell people in the 'South' that they should have fewer children, opening a Pandora's box of potential accusations (as discussed by Kopnina and Washington 2016).

The argument that 'population growth is not a problem' is simplistic, dividing people into the bad (rich, Western) consumers and the innocent (poor, non-Western) developing world. In fact, such oversimplification is really 'Reductio ad absurdum', where environmental impact is unrealistically divorced from the number of people (Kopnina and Washington 2016). Simplistic divisions also tend to underplay the impact of the growth of the middle classes in developing countries, and the environmental impact of the increasing population in poor countries (MEA 2005). The old mantra that the problem is just 'in the North' (i.e. the developed world) ignores the fact that the developing world is rapidly increasing its consumption (Washington 2015). Yet the world's planetary boundaries (thresholds) are already exceeded on four levels (climate change, nitrate pollution, phosphorus pollution and species extinction) and are close to exceeding thresholds for other boundaries (Rockstrom et al. 2009). Civilization cannot afford to stick with a fantasy that overpopulation is a non issue.

There is also the key problem that the political Left – and much of the environment movement – have blundered badly in terms of facing up to overpopulation as a key issue, as they have ignored or denied its essential importance (Crist 2012; Kopnina and Washington 2016). There is also the difficult and conflicted issue of *immigration* into nations in terms of population growth. In some nations, net immigration is responsible for over half the growth rate in population (such as Australia and the United States). Yet so conflicted and polarised has the immigration issue become that some strong population advocates (who have written important works on the issue) have now come out arguing that activists should talk about overpopulation but *not* immigration (Crist 2019). The reason put forward is that once that word 'immigration' is used, people switch off in regard to the whole topic of overpopulation. This is likely due to immigration being such a polarised issue, so that talk of reducing immigration puts people into a mental knot where it is easier to think of something else (so they effectively ignore the issue). As Palmer (2012: 103) notes:

> This puts many of us in a difficult spot. We do not want to deny the opportunity of immigration to other people. But we don't want to deny other opportunities, either, such as the opportunity for our children and grandchildren to have a healthy environment, or the opportunity or other species to live and thrive, and for our kids and grandchildren to see them.

Accordingly, immigration remains a most difficult issue, even for those who agree that overpopulation is a major problem. However, I believe we have an ecocentric 'duty of care' to look after nature in *every* nation. If nations are beyond their ecologically sustainable population (as Australia, the United States and much of Europe *are*) then ignoring immigration just leads to greater ecocide and extinction in those countries. Difficult though it may be to talk about immigration, its denial will lead to greater local ecocide, as we will not be able to stabilise the population in many countries who are growing in population mainly due to net immigration. Hence I believe the urgent need is to move to an ecologically sustainable population in *all* countries. This means we must talk about controlling immigration if that is pushing the population in a country beyond ecologically sustainable limits. Overpopulation beyond ecological limits is not just limited to 'other' countries; it is probably happening in your country as well. We need to act to stabilise population to an ecologically sustainable level in *every* country to protect the natural heritage of all countries.

Is talking about overpopulation anti-human?

The Discovery Institute has a video *The War on Humans*, which argues that any argument against population growth is anti-human (DI 2014). This is expanded on in the book by the same name (White 2014). They claim that anti-human activists want to reduce the human population by 90%. Clearly, they believe talking about overpopulation comes from a hatred of humanity (as discussed in Kopnina and Washington 2016). Of course, no serious evidence to support this is presented. Environmental scientists and scholars who point out the danger of overpopulation do so for two key reasons. The first is that this is causing ecocide and the extinction of life. The second is that the first reason is likely to lead to famine and war, and the *major loss of human population* that could be as high as a billion or more (this is no exaggeration). The loss of nonhuman life via massive extinction would be a tragedy. The loss of human life due to pushing ecosystems into collapse (including agro-ecosystems) would equally be a tragedy. Thus, talking about overpopulation is not anti-human but *very pro-human*. I have never met a population activist who wanted to see the death of hundreds of millions (perhaps billions) of people. Their population activism is precisely because they seek to avoid a human predicament where billions starve and die (human and nonhuman). Similarly, they wish to avoid a situation where international conflict and war are increased. The 'anti-human' epithet is thus an easy thing to say but has no evidence to support it. It is irrational. A moment's thought makes clear that on a finite planet human population cannot expand forever, especially when one passes beyond ecological limits, as we most certainly have (see Chapter 1). Breaking the denial dam about overpopulation is thus one of the most pro-human (as well as pro-nature) things any of us can do to heal the world.

Any discussion of balance, harmony, or planetary and human healing becomes impossible if population-control activists are pejoratively branded as 'rabid environmentalists', 'misanthropes', 'racists', or even betrayers of the human race (discussed by Kopnina and Washington 2016). Not only is such branding counter-productive, it reduces the very possibility of finding solutions to ensure a secure future for all the generations of tomorrow (human *and* nonhuman). It is not 'anti-human' to discuss this – on the contrary it shows the deepest concern for all future generations and the life they may have to endure.

Solutions: the big picture

Campbell (2012: 52) concludes we can break the spiral of silence about overpopulation by showing that it: 1) makes it impossible for the poor to escape poverty and ecological degradation; 2) high fertility is not due to women's desire to have more children; 3) fertility can decline when women are given freedom to control their fertility via family planning. Kopnina and Washington (2016: 139) conclude re overpopulation:

> We believe that addressing population is not a condemnation of the poor or an excusal of the rich. Nor is this a call for coerced population control or a perpetuation of social inequality. Rather, this is a call for recognition that there are many common factors, contributing to global poverty, inequality, and environmental destruction, and that population growth exacerbates all of these. Returning to the practicalities, we stress our belief that population control should never be coercive. Environmentalists value all life. We propose that humane and non-coercive but nonetheless urgent measures are considered seriously to prevent both social and environmental disaster.

They note that the first step is to accept that we (collectively) have a major problem. Then we need to abandon denial, and discuss and immediately implement solutions. Engelman (2016) shows that overpopulation can be tackled by nine humane (non-coercive) strategies to stabilise population:

1   Assure universal access to a range of safe and effective contraceptive options and family planning services for both sexes.
2   Guarantee education through secondary school for all, with a particular focus on girls.
3   Eradicate gender bias from law, economic opportunity, health and culture.

4   Offer age–appropriate sexuality education for all students.
5   End all policies that reward parents financially if they are based on the number of their children.
6   Integrate teaching about population, environment and development rela-tionships into school curricula at multiple levels.
7   Put prices on environmental costs and impacts.
8   Adjust to population aging rather than trying to delay it through govern-mental incentives or programmes aimed at boosting childbearing.
9   Convince leaders to commit to ending population growth through the exercise of human rights and human development.

BOX 4.2 HOW TO HELP REVERSE POPULATION GROWTH, ROBERT ENGELMAN

You are one of some 7.7 billion human beings alive on the Earth today. That may make you feel small and powerless when thinking about the over-whelming pressure of humans and their activities on the planet. But influen-cing population size is easier than you think. And it's more than a matter of just being aware that 'everyone counts'. That's not intended as a pun on demographic math. But it is indeed the case that one thing we can all do is learn more about population change and what influences it.

Remember the power of exponential growth. Any child born becomes a potential essential link to literally millions of descendants over many gener-ations. So if you are intentional about your own reproduction, and if you help others avoid pregnancies and births of children not intended, planned for or wanted by their parents, the beneficial demographic implications are unfathomably vast over long periods of time. Research demonstrates that there's really no other behavior in which you can engage with such powerful long-term benefit to the environment and the rest of life on earth.

So, let's start with asking whether you are in the age group where repro-duction is possible. If so, be intentional about sex and reproduction. That is, don't go for either without considerable forethought. Incredibly, an estimated two out of five pregnancies are unintended, and the proportion is actually higher in developed countries than in developing ones. That estimate is based on surveys that require women to assert that their current or recent preg-nancy was not planned. It's certainly conceivable – again, no pun intended – that the proportions are actually much higher. Either way, make sure that any pregnancy you have or are a partner to happens out of a desire to be a parent to a new human being, or perhaps two or more, for decades to come. This is probably the most important commitment you will make in your own life.

If you are not in the age of reproduction (or for any other reason you have no chance of pregnancy or partnership in one), there is still much you can do to help prevent unwanted pregnancies. Support sexual and repro-ductive health and rights, including the right to safe abortion. You can educate your own family, friends and other loved ones on the importance of healthy, intentional sexuality and reproduction. You can advocate that government at all levels facilitate removal of all barriers to the use of contraception and access to safe abortion. You can advocate for gender equality and an end to gender-based violence, so that no girl or woman anywhere will marry, have sex, or become pregnant against her will.

You don't have to stop with helping to prevent unintended pregnancy. Support the spread of comprehensive sexuality education and education generally. Nothing else is so powerful in reducing the desire to have large families. If enough of us combine all these steps, human population will soon retreat.

For more information:

Engelman, R. (2011) 'An end to population growth: Why family planning is key to a sustainable future', *Solutions Journal* **2** (3): 32–41, see: www.thesolutions journal.com/article/an-end-to-population-growth-why-family-planning-is-key-to-a-sustainable-future/.

Engelman, R. (2016) 'Nine population strategies to stop short of 9 billion', in *A Future beyond Growth: Towards a Steady State Economy*, eds. H. Washington and P. Twomey, New York: Routledge.

Such strategies do actually work. Using similar strategies, Iran was able to halve its population growth rate from 1987 to 1994 (Brown 2011). The Population Media Center (PMC n.d.) continues to successfully educate about such strategies. Using such strategies, we could reduce global population to 6 billion by the end of the century and to a sustainable 2–3 billion by the end of the following century (Staples and Cafaro 2012). However, to do this we must talk about it and break the denial dam that keeps overpopulation a taboo.

Conclusion

The denial of human overpopulation is a tragedy, both for nature and humanity. We are faced with three major drivers of unsustainability – overpopulation, overconsumption and the endless growth economy. Clearly, talking about any of these is not easy, as society is in denial of all of them (see Chapter 3). Nevertheless, we have to talk about *all of them*. We must see and act on all the elephants in the room, and that means that overpopulation can no longer be ignored or denied. Overpopulation is not the only key problem humanity faces; its terrible twin – overconsumption – is covered in Chapter 5 (and the growth economy is covered in Chapters 3 and 5). However, overpopulation and overconsumption are entwined and must be solved concurrently. While much of society and academia continue to ignore overpopulation as a key driver of unsustainability, however, any chance of being able to heal the world will remain vanishingly small.

What can *I* do?

There are various things each of us can do to foreground the issue of overpopulation:

1) *Don't ignore or deny the issue.* Discuss calmly what overpopulation means with friends and family – thus you can help to break the denial dam. Point out the reality that a constantly growing population is not good for either humanity or nature. Explain that not acting on overpopulation will bring great misery to both humanity and nature. Point out that acting on over-population is very *pro-human* not anti-human.
2) Explain that the human population can be controlled in humane and non-coercive strategies. We are *not* talking about euthanasia and mass sterilisa-tion, but about strategies such as effective (and adequately funded) family planning, and education for young women (Engelman 2016).
3) If the decision has not been made, consider having only one child (McKibben 1999). Two at most. If you want more you can always adopt.
4) Challenge the statement: 'It is a woman's right to have as many children as she wants'. Rights are not 'absolute' things, as future generations also have rights, as does nature herself (Washington 2018). Future generations and nature also deserve justice. Given the current world predicament, certainly it may be a woman's 'right' to have one child (perhaps two), but no more than that if we are to heal the world.
5) Consider pointing out politely to others who *plan* to have more than two children that, while they may argue 'Hey, we can support them!', the world today simply *cannot* (as we are in ecological overshoot).
6) Urge your government to establish an inquiry to determine what the eco-logically sustainable population of your nation is and to create strategies to move rapidly to this (through actions carried out by a specific statutory gov-ernment body). Most governments are very reluctant to consider this, so it will take a lot of political lobbying to get action. For example, in Australia, various estimates for an ecologically sustainable population have been devel-oped that have ranged from 10 to 12 to 15 million (Washington 2015), with the highest being from 'Population 2040: Australia's Choice' run by the Australian Academy of Sciences (AS 1994) – which was *23 million*. However, Australia's population is now over 25 million (with most growth coming from net migration). No government in Australia has a strategy aiming to reach an ecologically sustainable population, however.
7) Talk about the equation Environmental Impact = Population × Affluence × Technology, where society needs to *work on all three*. Talk about how the Global Ecological Footprint is at least 1.7 Earth's (yet we have only one). Explain that we need action on both overpopulation and overconsumption. Explain that ignoring either just promotes ongoing ecocide and greater social problems.

8) Urge your government to promote effective (and properly resourced) *family planning* that makes birth control technologies readily available to all, not the 'women's reproductive health' the UN (and others) now speaks of (which commonly doesn't actually promote family planning, Campbell 2012). If you are in a country that provides overseas aid, urge that some of this be used to support family planning overseas.

References

Abegao, J. (2018) *Human Overpopulation Atlas: Volume 1*, see: www.overpopulationatlas.com/.

AS. (1994) 'Population 2040: Australia's choice', see: www.science.org.au/news-and-events/news-and-media-releases/scientists-pose-solutions-big-australia (accessed 27 March 2019).

Assadourian, E. (2013) 'Re-engineering cultures to create a sustainable civilization', in *State of the World 2013: Is Sustainability Still Possible?*, ed. L. Starke, Washington: Island Press, pp. 113–125.

BBC. (2006) 'Earth is too crowded for Utopia', *BBC News Online*, 6 January 2006, see: http://news.bbc.co.uk/2/hi/science/nature/4584572.stm (accessed 27 March 2019).

Beck, R. and Kolankiewicz, L. (2001) *Forsaking Fundamentals: The Environmental Establishment Abandons US Population Stabilisation*, Washington: US Centre for Immigration Studies, see: www.cis.org/articles/2001/forsaking/toc.html (accessed 27 March 2019)

Blowfield, M. (2013) *Business and Sustainability*, Oxford, UK: Oxford University Press.

Braungart, M. and McDonough, W. (2008) *Cradle to Cradle: Remaking the Way We Make Things*, London: Vintage Books.

Brown, L. (2011) *World on the Edge: How to Prevent Environmental and Economic Collapse*, New York: W.W. Norton and Co.

Brown, L. (2012) *Full Planet, Empty Plates: The New Geopolitics of Food Scarcity*, New York: Norton.

Butler, T. (2012) 'Colossus versus liberty: A bloated humanity's assault on freedom', in *Life on the Brink: Environmentalists Confront Overpopulation*, eds. P. Cafaro and E. Crist, Georgia, USA: University of Georgia Press, pp. 160–171.

Cafaro, P. and Crist, E. (2012) *Life on the Brink: Environmentalists Confront Overpopulation*, Georgia, USA: University of Georgia Press.

Campbell, M. (2012) 'Why the silence on population?', in *Life on the Brink: Environmentalists Confront Overpopulation*, eds. P. Cafaro and E. Crist, Georgia, USA: University of Georgia Press, pp. 41–55.

Carrington, D. (2018) 'Paul Ehrlich: "Collapse of civilisation is a near certainty within decades"', *The Guardian*, 22 March, see: www.theguardian.com/cities/2018/mar/22/collapse-civilisation-near-certain-decades-population-bomb-paul-ehrlich (accessed 27 March 2019).

Catton, W. (1982) *Overshoot: The Ecological Basis of Revolutionary Change*, Chicago: University of Illinois Press.

Collins, J. and Page, L. (2019) 'The heritability of fertility makes world population stabilization unlikely in the foreseeable future', *Evolution and Human Behaviour*, **40** (1): 105–111.

Collins, P. (2010) *Judgment Day: The Struggle for Life on Earth*, Sydney: UNSW Press.

Crist, E. (2012) 'Abundant Earth and the population question', in *Life on the Brink: Environmentalists Confront Overpopulation*, eds. P. Cafaro and E. Crist, Georgia: University of Georgia Press, pp. 141–151.

Crist, E. (2019) 'Decoupling the global population problem from immigration issues', *The Ecological Citizen*, **2** (2): 149–151.

Crist, E., Mora, C. and Engelman, R. (2017) 'The interaction of human population, food production, and biodiversity protection', *Science*, **356**: 260–264.

Derer, P. (2018) 'Evidence for the changing discourse on population growth in an environmental magazine', The Overpopulation Project, 16 Aug 2018, see: https://overpopulation-project.com/2018/08/16/evidence-for-the-changing-discourse-on-population-growth-in-an-environmental-magazine/.

DI. (2014) *The War on Human Video*, Discovery Institute, see: www.youtube.com/watch?v=RWcEYYj_-rg (accessed 15 March 2019).

Dietz, R. and O'Neill, D. (2013) *Enough Is Enough: Building a Sustainable Economy Is a World of Finite Resources*, San Francisco: Berrett-Koehler Publishers.

Ehrlich, P. (1968) *The Population Bomb*, New York: Ballentine Books.

Ehrlich, P. (2013) *Personal Communication from Prof. Paul Ehrlich at the 2013 Fenner Conference on the Environment 'Population, Resources and Climate Change'*, Canberra Australia, see: http://population.org.au/fenner-conference-2013-summing (accessed 27 March 2019).

Ehrlich, P. (2017) 'You don't need a scientist to know what's causing the sixth mass extinction', *The Guardian*, 12 July 2017, see: www.theguardian.com/commentisfree/2017/jul/11/sixth-mass-extinction-habitats-destroy-population (accessed 20 March 2019).

Ehrlich, P., Ehrlich, A. and Holdren, J. (1977) *Ecoscience: Population, Resources, Environment*, New York: WH Freeman and Co.

Engelman, R. (2012) 'Nine population strategies to stop short of 9 billion', in *State of the World 2012: Moving toward Sustainable Prosperity*, ed. L. Starke, Washington: Island Press, pp. 121–128.

Engelman, R. (2013) 'Beyond sustainababble', in *State of the World 2013: Is Sustainability Still Possible?*, ed. L. Starke, Washington: Island Press, pp. 3–16.

Engelman, R. (2016) 'Nine population strategies to stop short of 9 billion', in *A Future beyond Growth: Towards A Steady State Economy*, eds. H. Washington and P. Twomey, New York: Routledge, pp. 32–42.

Erb, K., Haberl, H., Krausmann, F. et al. (2009) 'Eating the planet: Feeding and fueling the world sustainably, fairly and humanely – A scoping study', Social ecology working paper no. 116, Vienna: Institute of Social Ecology and Potsdam Institute for Climate Impact Research.

FAO. (2011) *The State of the World's Land and Water Resources for Food and Agriculture (SOLAW)*, Rome: Food and Agriculture Organisation.

Flavin, C. (2010) 'Preface', in *State of the World 2010: Transforming Cultures from Consumerism to Sustainability*, eds. L. Starke and L. Mastny, New York: Worldwatch Institute/Earthscan, pp. xvii–xx.

Fletcher, R. (2014) 'World population to hit 11bn in 2100 – With 70% chance of continuous rise', E-mail posted on Environmental Anthropology Listserv (EANTHL@LIST SERV.UGA.EDU), 2014 Sept 19.

Fletcher, R., Breitlin, J. and Puleo, V. (2014) 'Barbarian hordes: The overpopulation scapegoat in international development discourse', *Third World Quarterly*, **35**: 1195–1215.

Hartmann, B. (2004) 'Conserving racism: The greening of hate at home and abroad', see: www.researchgate.net/publication/265045644_Conserving_Racism_The_Greening_of_Hate_at_Home_and_Abroad (Accessed 27 March 2019).

Hartmann, B., Meyerson, F., Guillebaud, J., Chamie, J. and Desvaux, M. (2008) 'Population and climate change', *Bulletin of Atomic Scientists*, 16th April 2008.

Hulme, M. (2009) *Why We Disagree about Climate Change: Understanding Controversy, Inaction and Opportunity*, Cambridge: Cambridge University Press.

IPBES. (2019) 'Nature's dangerous decline "unprecedented" species extinction rates "accelerating"', Press release Intergovernmental Science-Policy Platform on Biodiversity and Ecosystem Services, see: www.ipbes.net/news/Media-Release-Global-Assessment.

IPCC. (2014) *Climate Report 2014: Synthesis Report. International Panel on Climate Change*, see: https://archive.ipcc.ch/pdf/assessment-report/ar5/syr/SYR_AR5_FINAL_full_w cover.pdf. (accessed 20 March 2019).

Kallis, G. (2018) *Degrowth*, New York: Columbia University Press.

Kopnina, H. and Washington, H. (2016) 'Discussing why population growth is still ignored or denied', *Chinese Journal of Population Resources and Environment*, **14** (2): 133–143.

McKibben, B. (1999) *Maybe One*, London: Anchor.

MEA. (2005) *Living beyond Our Means: Natural Assets and Human Wellbeing, Statement from the Board, Millennium Ecosystem Assessment*, United Nations Environment Programme (UNEP), see: www.millenniumassessment.org/documents/document.429.aspx.pdf (accessed 19 February 2018).

Monbiot, G. (2009) 'The population myth', see:www.monbiot.com/archives/2009/09/ 29/the-population-myth/ (accessed 27 March 2019).

Palmer, T. (2012) 'Beyond futility', in *Life on the Brink: Environmentalists Confront Overpopulation*, eds. P. Cafaro and E. Crist, Georgia: University of Georgia Press, pp. 98–107.

PM. (2010) 'Capacity population', Population Matters leaflet in 2010 now no longer on their website, see: https://populationmatters.org/.

PMC. (n.d.) 'Population Media Center website', see: www.populationmedia.org/.

Raworth, K. (2017) *Doughnut Economics*, London: Cornerstone.

Rees, W. (2019) 'End game: The economy as eco-catastrophe and what needs to change', *Real World Economic Review*, **87**: 132–148.

Robbins, P. (2012) *Political Ecology: A Critical Introduction*, 2nd edition, Malden, MA: Wiley.

Rockstrom, J., Steffen, W., Noone, K. et al. (2009) 'Planetary boundaries: Exploring the safe operating space for humanity', *Ecology and Society*, **14** (2): 32, see: www .ecologyandsociety.org/vol14/iss2/art32/ (accessed 19 March 2019).

Smail, K. (2003) 'Remembering Malthus III: Implementing a global population reduction', *American Journal of Physical Anthropology*, **122**: 295–300.

Smail, K. (2016) 'Excessive human numbers in a world of finite limits: Confronting the threshold of collapse', in *Handbook of Environmental Anthropology*, ed. S. O. Kopnina, London, UK: Routledge, pp. 384–398.

Staples, W. and Cafaro, P. (2012) 'For a species right to exist', in *Life on the Brink: Environmentalists Confront Overpopulation*, eds. P. Cafaro and E. Crist, Georgia, USA: University of Georgia Press, pp. 283–300.

Steffen, W., Richardson, K., Rockström, J. et al. (2015) 'Planetary boundaries: Guiding human development on a changing planet', *Science*, **347** (6223), DOI: 10.1126/ science.1259855.

The Economist. (2012b) 'America's demographic squeeze: Double bind', see: www.econo mist.com/united-states/2012/12/15/double-bind (accessed 27 March 2019).

The Economist. (2012a) 'Indonesia's forests and REDD: Palming off', see: www.economist .com/blogs/banyan/2012/12/indonesias-forests-and-redd (accessed 27 March 2019).

UCS. (2018) 'Each country's share of CO_2 emissions', Union of Concerned Scientists, see: www.ucsusa.org/global-warming/science-and-impacts/science/each-countrys-share-of-co2.html (accessed 12 March 2019).

UNDESA. (2017) *World Population Projected to Reach 9.8 Billion in 2050, and 11.2 Billion in 2100*. United Nations Department of Economic and Social Affairs, see:www.un.org

/development/desa/en/news/population/world-population-prospects-2017 .html (accessed 26 August 2018).

Washington, H. (1991) *Ecosolutions: Solving Environmental Problems for the World and Australia*, Tea Gardens, NSW, Australia: Boobook Publications.

Washington, H. (2015) *Demystifying Sustainability: Towards Real Solutions*, London: Routledge.

Washington, H. (2018) *A Sense of Wonder Towards Nature: Healing the World through Belonging*, London: Routledge.

Washington, H. (2019) Unpublished poem 'enough'.

Weeden, D. and Palomba, C. (2012) 'A post-Cairo paradigm: Both numbers and women matter', in *Life on the Brink: Environmentalists Confront Overpopulation*, eds. P. Cafaro and E. Crist, Athens: University of Georgia Press, pp. 255–273.

Weld, M. (2012) 'Deconstructing the dangerous dogma of denial: The feminist-environmental justice movement and its flight from overpopulation', *Ethics in Science and Environmental Politics*, **12**: 53–58.

White, R. (1994) 'Green politics and the question of population', *Journal of Australian Studies*, **18**: 27–43.

White, W. (2014) *The War on Humans*, Seattle, WA: Discovery Institute Press.

WPB. (n.d.) 'Current population is three times the sustainable level', see: www. worldpopu lationbalance.org/3_times_sustainable (accessed 17 March 2019).

WWF. (2018) *Living Planet Report – 2018: Aiming Higher*, World Wide Fund for Nature, eds. M. Grooten and R. E. A. Almond, Gland, Switzerland: WWF.

5

TRANSFORMING SOCIETY'S ASSUMPTIONS

What can I say
Of this absurdity?
This commitment to 'ever more'
This psychotic delusion,
And denial of reality.

From 'Endless' (Washington 2018)

SUMMARY

Western society (now globalised around the world) is bedevilled by unsustainable assumptions that hinder healing the world. These are either not true, or make things far worse. Ten such assumptions are identified, seven of which have been covered in other chapters. The three problem assumptions discussed in this chapter are the assumptions of neoclassical economics, consumerism and what can be called 'assumptions of hate' (where some in society spend a lot of energy hating someone or something). For the detail explaining this summary, please read on; however, if you just want to know the things *you can do* (30 are listed) about this, then turn to the end of the chapter.

Introduction

Our society is bedevilled by 'assumptions'. We *assume* many things that are either not true or hinder us healing the world. The old joke holds true that if you *assume* – you make an 'ass' out of 'you' and 'me'. It can be seen below that most of these are related to society's worldview and ethics (see Chapter 2). Many of our assumptions hold us back from healing the world. Such problem assumptions include:

1. Modernism (Oelschlaeger 1991) (see Chapter 2)
2. Anthropocentrism ('Sole Value' and 'Greater Value' assumptions, Curry 2011) (see Chapter 2)
3. Growthism and evermoreism (= 'no limits') (see Chapter 3)
4. Resourcism and infinite resources (see Chapter 3)
5. Optimism (it's all fine!) or pessimism (we are all doomed!) (see Chapter 10)
6. Ownership and human mastery (see Chapter 8)
7. Techno-centrism and ecomodernism (see Chapter 6)
8. The assumptions of neoclassical economics
9. Consumerism
10. Assumptions of hate.

Clearly there are a lot of assumptions that act to stop us healing our world. I am not suggesting these are the only false assumptions in society, simply ten key assumptions that hinder healing. The first seven assumptions are covered elsewhere in this book. There remain three other key assumptions to discuss that hold us back from healing the world.

The assumptions of neoclassical economics

There are assumptions that neoclassical economics makes about how the world works (Diesendorf and Hamilton 1997; Costanza et al. 2013). These should (indeed must) be examined if we are to heal our world. Rees (2019: 134) notes that: 'all economic theories/paradigms are elaborate conjectures and that none can contain more than a partial representation of biophysical, or even social, reality'. Washington (2015) has summarised the assumptions of neoclassical economics from several sources, the most eminent being renowned ecological economist Herman Daly:

1. Strong *anthropocentrism*. Nature is seen as 'just a resource' to be used to provide the greatest 'utility' to the greatest number of people. Rolston (2012: 37) notes that if neoclassical economics is the 'driver we will seek' for our society, then it will result in 'maximum harvests in a bioindustrial world', as the current economic model is extractive in nature and commodifies the land. Land becomes merely 'resources' and 'natural capital' (see Chapter 3). Such maximum harvests will not consider the limits (or tipping points) of ecosystems (see Chapter 10).
2. The idea that the *free market* will control all that is needed, that the 'invisible hand' will regulate things for human benefit (Daly 1991). This is a key 'given truth' of neoliberalism and has become almost a religion (Daly 2008).

The deification of the free market has also been shown to be an underlying ideological reason for conservatives to deny climate change (Oreskes and Conway 2010). Stiglitz (2002) argues the invisible hand is 'invisible' because it is not there. Common and Stagl (2005) note the invisible hand does not in fact work. Market failures of various kinds mean that actual market outcomes are not efficient. Rees (2010: 10) notes that: 'Unsustainability may be the greatest example of market failure'. Renner and Prugh (2014) note that markets don't have a social conscience, environmental ethic or long-term vision. Neither do they conserve the 'public good' (Wijkman and Rockstrom 2012). Achieving efficiency does not guarantee equity, between either those alive at one point in time, or different points in time. Accordingly, it doesn't consider inter-generational equity (Washington 2015). Even under 'ideal' conditions, market outcomes may be very unfair (especially to future generations and the nonhuman world).

3.   The idea that the economy can *grow forever* in terms of continually rising GDP, which increased by an astounding 25-fold over the last century (Dietz and O'Neill 2013). Daly (1991: 8) notes that economic growth is the most 'universally accepted goal' in the world and that: 'Capitalists, communists, fascists and socialists all want economic growth and strive to maximise it' (see also Chapter 3).

4.   The *refusal to accept any biophysical limits to growth*, because when classical economics was developed, limits were distant (Daly 1991). However, neoclassical economics today still does not acknowledge any limits on a finite Earth, and most economists are not taught how nature actually works (Wijkman and Rockstrom 2012). Daly (1991) notes that three interrelated conditions – finitude, entropy and complex ecological interdependence – combine to provide the biophysical limits to growth.

5.   A *circular theory of production* causing consumption that causes production in a never-ending cycle (as shown in Figure 5.1). Daly (1991) notes that real production and consumption are in no way circular. The growth economy sees outputs returned as fresh inputs, and he notes ironically this requires we: 'discover the secret of perpetual motion' (Ibid.: 197). 'Money fetishism' is the idea that money flows in an isolated circle, and thus so can commodities. Daly and Cobb (1994: 37) argue the neoclassical view believes that if money balances can grow bigger forever at compound interest, then so can real GNP, and they respond ironically: 'so can pigs and cars and haircuts'. The idea of a circular theory of production is a classic case of the 'Fallacy of misplaced concreteness', which Georgescu-Roegen (1971) thought was the key sin of standard economics. Philosopher Alfred Whitehead (1929: 11) defines this as: 'neglecting the degree of abstraction involved when an actual entity is considered, merely so far as it exemplified certain categories of thought'. In other words, reductionism and modelling pose the risk that the model becomes 'more real' than reality. The abstractions necessary to create mechanistic models: 'always do violence to reality' (Daly 1991: 46).

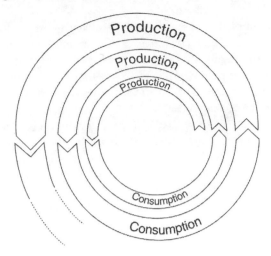

FIGURE 5.1   The assumed 'circular flow' of production and consumption in the neoclassical economy (Source: adapted from figure 7 in Daly (1991) *Steady State Economics*).

Policy decisions are: 'determined by mathematical theorems whose virtue is their deductive fruitfulness rather than their connection to the real world' (Daly and Cobb 1994: 96). The abstraction has gone too far and the practitioners are not aware of this, for: 'the fallacy of misplaced concreteness is too pervasive' (Daly and Cobb 1994: 96). Those who habitually think in terms of abstract reductionist models are especially prone to this fallacy. Mathematical models and computer simulations are the hallmark of neoclassical economics. Such elaborate logical structures prize theory over fact and reinterpret fact to fit theory (Daly and Cobb 1994).

6.   Neoclassical economics *ignores the Second Law of Thermodynamics* and fails to consider 'entropy' as a key feature of economics and reality. Georgescu-Roegen (1971) and Daly (1991) explore this issue. Thermodynamics shows that we do not create or destroy anything in a physical sense, we merely transform or rearrange it. The inevitable cost of arranging greater order in one part of the system (the human economy) is to create disorder elsewhere – nature (Daly 1991). 'Entropy' is a measure of the disorder in a closed system. In thermodynamics, low entropy quantities (usable energy, high quality raw materials) move to high entropy quantities (waste heat and wastes). Entropy is the basic physical coordinate of 'scarcity'. Were it not for entropy, we could burn the same gallon of petrol over and over, and our capital stock would never wear out. Technology cannot rise above the laws of physics, so there is no question of 'inventing a way to recycle energy', even if some economists seem to imply this (as discussed by Daly 1991: 36). Economic modelling cannot ignore entropy, otherwise it becomes irrelevant to reality. Such physical laws will not go away, and Einstein thought the entropy law the: 'least likely to ever be overthrown' (in Daly 1991: 36).

7. Environmental damage is *merely an 'externality'*. The spillover effects of market transactions have been named 'externalities'. Externalities are costs or benefits arising from an economic activity that affect somebody other than the people engaged in it and are not reflected fully in prices. Environmental damage is known as a 'negative externality', something external to the economic model. An externality (to neoclassical economists) is seen as being worth only peripheral attention (Daly and Cobb 1994). This is a key part of what has led to the environmental crisis. Of course, environmental crises can still occur even where externalities *are* 'internalised' (incorporated into market accounting). Foxon et al. (2012) point out that this approach is inadequate for solving climate change and biodiversity loss (the pricing of 'natural capital' and 'ecosystem services' highlights this, Washington 2019).

8. All forms of *capital can be substituted*, where the assumption is that human capital can be substituted for natural capital (the claim of 'weak sustainability'). This fundamentally denies the reality that society is fully dependent on nature, hence 'weak sustainability' is not in fact sustainable at all (Washington 2015).

The above assumptions have been detailed individually by others, principally Herman Daly (Daly 1991, 1996, 2014; Daly and Cobb 1994), though rarely brought together in one place as shown here. These assumptions show the fundamental challenge of our current neoclassical economics is that it doesn't function within ecological and social realities. In fact neoclassical economics is optimally badly structured if we want to live sustainably or heal the world. As Rees (2019) accurately notes, we have a: 'global economic system whose conceptual framing, operating assumptions and *de facto* practices are pathologically incompatible with the very ecosystems that sustain it'. Looking at these assumptions from the viewpoint of environmental science (my field of research), the above assumptions are bizarre and unsustainable. However, they underpin the reigning neoclassical economic synthesis to this day (Washington 2015). The above assumptions deserved to be challenged, and some economists and scientists have been doing so since the 1970s (e.g. Daly 1991, 1996, 2014; Victor 2008, 2019; Costanza et al. 2013; Czech 2013; Dietz and O'Neill 2013; Washington and Twomey 2016).

BOX 5.1 THE FUTURE OF ECOLOGICAL ECONOMICS, JOSHUA FARLEY

Mainstream economists have defined the economic problem as the allocation of resources in conditions of scarcity: how do we allocate finite economic resources of labor, capital, energy and raw materials to best satisfy the limitless wants and needs of inherently self-interested humans? Their answer is competitive markets that channel self-interest and greed towards the maximization of monetary value. Infinite human ingenuity and the profit motive will drive the innovation of new resources if we exhaust existing ones and of substitutes for the essential ecological functions we destroy in the process. Our insatiability demands endless economic growth. Harm to other species only matters if it reduces human welfare. This mindset has led to unprecedented destruction of global ecosystems and biodiversity as well as dire threats to human civilization and even survival.

Ecological economics arose in response to these challenges. It recognizes that the human system is an integral part of our finite global ecosystem. It is impossible to make something from nothing, do work without energy, or make nothing from something. We use fossil fuels to physically transform raw materials from nature into economic products and waste spewed back into the ecosystem. The raw materials we convert alternatively serve as the structural building blocks of ecosystems: when we remove them and dump our wastes, we degrade the capacity of ecosystems to generate ecological functions essential to the survival of humans and all other species. The economic problem is how to allocate the planet's finite resources between economic activities essential to human welfare and conservation to sustain the non-substitutable ecological functions upon which humans and the rest of nature depend.

The most serious ecological challenges we face are social dilemmas, 'a situation in which members of a group can gain by cooperating, but cooperation is costly, so each individual does better personally by not cooperating, no matter what the others do' (Gintis 2011). Competitive self-interest cannot solve these problems. An economic transition is required. So too is an energy transition, which serves to illustrate our problem. Our current economy is powered by fossil fuels. If I burn a barrel of oil, you cannot. Competition is unavoidable and private property rights ubiquitous. Aside from the existential threats they pose to humanity (!), fossil fuels are a good fit for the market economy. We must transition to alternative energy. No matter how much solar energy one country captures, it has no impact on the amount available to others. Technologies for capturing solar energy improve through use. There is no need for competition. Private property rights to green technologies incentivize market production, but may price them out of the reach of poorer countries, who will continue to burn coal, making everyone worse off. It is not even possible to create private property rights to a stable climate, healthy ozone layer, or most ecological functions. Collective choice and cooperation are the only solutions. Humans are not inherently self-interested, but rather the most cooperative species to ever evolve (Wilson 2012). The future of ecological economics lies in recognizing that we should not seek competitive solutions to cooperation problems. Cooperation must define economic interactions with other humans as well as our interactions with the rest of nature. In the words of evolutionary biologist D.S. Wilson, 'the benefits of cooperation are like money in the bank earning compound interest' (Wilson et al. 2013).

For more information:

https://www.uvm.edu/gund
Gintis, H. (2011) 'Gene-culture coevolution and the nature of human sociality'. *Philosophical Transactions of the Royal Society of London. Series B, Biological Sciences*, **366**: 878–888.
Wilson, D. S., Ostrom, E., Cox, M. E. (2013) 'Generalizing the core design principles for the efficacy of groups'. *Journal of Economic Behavior & Organization*, **90** (Suppl): S21–S32.
Wilson, E.O. (2012) *The Social Conquest of Earth*, New York: Liveright Publishing Corporation.

Consumerism

Too often 'consumerism' is regarded as if it reflected a 'fundamental human nature' of material self-interest and possessiveness. This could hardly be further from the truth (Washington 2015). Our almost neurotic need to shop and consume is instead a reflection of how 'deeply social' we are. Living in unequal and individualistic societies, we use possessions to show ourselves in a good light and avoid appearing incompetent and inadequate (Wilkinson and Pickett 2010). If an important part of consumerism is driven by status competition, this explains why we continue to pursue economic growth *despite* its apparent lack of benefits. The growth of consumerism and the weakening of community life are related (Ibid.). We effectively allow a small social elite to hoard a greater percentage of resources. The 'Gini' coefficient measures the distribution of wealth, where 0 is where everyone has an equal share and 100 is where one person has all the wealth. The Gini coefficient has been rising recently in many countries. The United States is 41 while China is 42 and South Africa is 63 (World Bank, n.d.). In contrast it is 27 in Finland and Norway, 29 in Sweden, 32 in Japan and 33 in the UK (Ibid.).

Humanity's Indigenous cultures in the past (and today still in places) had largely worked out how to manage 'the commons' sustainably (Ostrom 1990). However, modern humanity has been plagued by the 'Tragedy of the Commons' (Hardin 1968), which one can describe as free access and unrestricted demand for a finite resource that ultimately depletes it. Hence modern society has overgrazed pasture, overlogged forests and overfished fisheries. Similarly, we are now mining resources and creating pollution far beyond what natural systems can detoxify. In effect, we in the West are told by constant advertising we need to live like princes and princesses, and our lifestyles and our consumption are on a violent collision course with nature (Higgs 2014). Hence we cannot continue to live as we have (Wijkman and Rockstrom 2012). As Paul Ekins (1991) has noted, a *sustainable* 'consumer society' is a contradiction in terms. Even in the 1950s, retail analyst Victor Lebow (1955: 7) concluded:

> Our enormously productive economy demands that we make consump-
> tion our way of life, that we convert the buying and use of goods into
> rituals, that we seek our spiritual satisfaction and our ego satisfaction in
> consumption. ... We need things consumed, burned up, worn out,
> replaced and discarded at an ever-increasing rate.

For some people today, consumption has become the meaning of life, the 'chief
sacred' (Ellul 1975). Yet the same consumption that has lately become the
'meaning of life' for some is now revealed as a barrier to healing the world.
Ultimately, we cannot roll back denial of the environmental crisis unless we roll
back our consumer worldview (Starke and Mastny 2010). Consumer cultures
exaggerate the forces that have allowed human societies to supposedly 'outgrow'
their environmental support systems (Assadourian 2010). We have not in fact
outgrown the ecosystem services that support us (and simply cannot). Humanity
remains completely dependent on nature (Cardinale et al. 2012; Washington
2013). Hamilton (2010) argues that many of us have constructed a 'personal
identity' through shopping and consumerism. We have substituted consumerism
for meaning (Collins 2010). Asking people embedded in the consumer myth to
change their consumerism may thus be like asking them to change their identity.
The issue is further complicated because the oft-stated aim to 'end global poverty' has become synonymous with the spread of a high consumption standard
of living (Crist 2012: 7). However, it is time to openly discuss the urgent need
to alter today's consumer consciousness. Consumerism and overconsumption are
not the way to heal our world.

Sukhdev (2010) notes we have become attuned to giving 'yes' answers for
trade-off choices that result in more consumption, more private wealth, more
physical capital – as against *better* consumption, more public wealth, more natural
capital. Society extends this false logic implicitly when we ignore the depletion
of forests, fisheries and so forth. This is part of a perverted logic of 'promoting
growth' or 'promoting development', without defining these terms in holistic or
equitable ways (from either an intra- or inter-generational perspective) (Ibid.).
While we promote endless growth on a finite planet, mainstream society doesn't
think about whether this is fair to future generations, or indeed fair to the rest
of life we share this planet with. Tacey (2000: 185) points out that:

> The consumerist society reinforces our shrunken empty status. It is vitally
> important for capitalism that we continue to experience ourselves as
> empty and small, since this provides us with the desire to expand and
> grow, and this desire is what consumerism is based on. Consumerism
> assumes that we are empty but permanently unable to fulfil our spiritual
> urge to expand. It steps into the vacuum and offers its own version of
> expansion and belonging ... If we stopped believing in the myth of our
> shrunken identity, the monster of consumerism would die, because it
> would no longer be nourished by our unrealized spiritual urges.

Mass consumption requires that consumer demand remains insatiable (Westra 2008). Preventing the collapse of human civilisation, however, requires nothing less than a wholesale transformation of dominant Western consumer culture (Flavin 2010). Consumption has gone up sixfold since 1960, but numbers have only grown by a factor of 2.2. Consumer expenditure per person has almost tripled (Assadourian 2010). According to the Global Footprint Network, humanity now uses the resources and services of 1.7 Earth's (GFN 2019), an unsustainable situation. If all the world were to adopt US (or Australian) lifestyles, we would need at least four more planets to supply them (Graff 2010). Assadourian (2010) suggests three goals to tackle consumerism. Firstly, consumption that undermines well-being has to be discouraged. Secondly, we need to replace private consumption of goods with public consumption of services (e.g. libraries, public transport). Thirdly, necessary goods must be designed to last and be 'cradle to cradle' recyclable. Wilkinson and Pickett (2010: 226) point out also that if we improve 'equality of income' in our societies, then consumer pressure will decline. Rees (2010) argues that if we are to reduce the human ecological footprint:

> [t]he fetishistic emphasis in free-market capitalist societies on individualism, competition, greed, and accumulation must be replaced by a renewed sense of community, cooperative relationships, generosity, and a sense of sufficiency; short-term material wants must give way to long-term basic needs.

The consumer ethic, seen as 'natural' by many consumers, is actually a *cultural teaching*, a purposeful social construct (Assadourian 2010). Following World War II, the United States was 'blessed' with great industrial capacity and large numbers of under-employed workers (returned soldiers). To take advantage of this abundant labour and break people out of their wartime habit of thriftiness, industry organised to legitimise profligate consumption, to make it a 'spiritual activity' (Rees 2008). In fact, many people resisted the throwaway society when it was first promulgated, as they believed in *thriftiness*. Three sectors aided the spread of consumerism: the car industry, the pet industry and the fast food industry (Assadourian 2013). The notion of 'perpetual growth' in consumption is thus a social construct, one vastly escalated as a transition strategy to reboot the economy after World War II. It has now (well and truly) run its useful course. However, what society has constructed it can theoretically decon-struct and replace (Washington 2015). The time has come for a new social contract that recognises humanity's collective interest in designing a better form of prosperity for a world with ecological limits (Moore and Rees 2013). This is a challenge, but it is also an opportunity to get things right so we can heal the world.

Consumer cultures will have to be reinvented as a culture of sustainability, so that living sustainably as an 'ecological citizen' who seeks to heal the world, feels as natural as living as a consumer does today (Assadourian 2013: 113). It will not be easy and will be forcefully resisted by many interests, such as the fossil fuel industry, big agribusiness, food processors, the fast food industry, car manufacturers, advertisers, etc. (Washington 2015). However, either we find ways to wrestle our cultural patterns out of the grip of vested interests, or Earth's eco-systems will decline further and bring down the consumer culture in a much crueller way (Assadourian 2013). To break free of consumerism, we will need to use all our social institutions: business, media, marketing, government, education, social movements and social traditions (Ibid.). We will need to engage the public in a conversation about the growing and unsustainable costs of the consumer society (Ibid.). 'Choice editing', or strategies for sustainability where society bans some 'high impact' products, will be needed (Ibid.). Consumer choice can help eliminate unsustainable choices (Kopnina and Blewitt 2015).

The 'finger of blame' can validly be pointed directly to the main agent of the 'brown' economy: *corporations* and the rules that govern their operations (Sukhdev 2013). Corporations are responsible for 60% of world GDP (Prugh 2013). Better policies are needed regarding accounting practice, taxation, financial leverage and advertising. These could result in a 'new corporate model', an agent for tomorrow's sustainable economy. We need:

1. Tax shifting to tax the 'bads'.
2. Rules and limits to govern financial leverage.
3. Advertising norms and standards that make it more responsible and accountable.
4. All major corporate externalities (positive or negative) to be measured, audited and reported in annual statements (Sukhdev 2013: 144).

We also need to ask what drives today's unsustainable consumption? It is no exaggeration to say that corporate advertising is the biggest single force driving consumer demand today. Daly (2008) suggests an 'Advertising Tax' to reduce advertising pressure. Victor (2008) argues advertising must now return to strictly factual information. Sukhdev (2013: 150) suggests four strategies for more accountable advertising:

1. Disclose lifespan of products in all ads.
2. Disclose countries of origin.
3. Recommend on the product 'how to dispose of it'.
4. Voluntarily submit a 10% development donation on total advertising spent in developing countries to support local sustainability projects.

BOX 5.2 VOLUNTARY SIMPLICITY: REIMAGINING THE GOOD LIFE BEYOND CONSUMER CULTURE, SAMUEL ALEXANDER

Increasingly it is recognised that human economic activity is breaching planetary boundaries and exceeding sustainable limits to growth. There is no way all 7.7 billion people on Earth could consume resources like we do in the most developed regions of the world, especially when the global population is trending toward 10 billion by mid-century.

Furthermore, technology and free markets alone are not going to produce a sustainable way of life. We've been trying those strategies for decades, and our ecological impacts are only intensifying as efficiency gains in production are reinvested in more growth and more consumption, not reduced impacts. Any resolution to this 'limits to growth' predicament is necessarily going to involve high-impact consumers learning how to live well on far fewer resources. This is not only a deep cultural challenge. We must also create systems and structures that support lifestyles of material sufficiency, not material affluence.

Fortunately there is growing movement (or movements) of people exploring alternatives to consumerism, marching under banners like voluntary simplicity, downshifting, permaculture, and minimalism. These movements can be broadly understood to be made up of people who are resisting high-consumption lifestyles and who are seeking, in various ways, a lower consumption but higher quality of life alternative.

This living strategy typically involves transferring progressively more of one's time and energy away from materialistic sources of satisfaction (e.g. money, assets, possessions, etc.) toward non-materialistic sources of satisfaction (e.g. social relations, community or civic engagement, creative activity, home-based production, self-development, spiritual exploration, relaxation, etc.).

While the practices and values of voluntary simplicity take many forms, prominent simplicity theorist, David Shi (2007), has suggested that some of the primary attributes of the movement include: thoughtful frugality; minimising expenditure on consumer goods and services; a reverence and respect for nature (and its limits); a desire for increased self-sufficiency; a commitment to conscientious rather than conspicuous consumption; a privileging of creativity and contemplation over possessions; an aesthetic preference for minimalism and functionality; and a sense of responsibility for the just uses of the world's resources. In the words commonly attributed to Gandhi: 'Live simply so that others may simply live'.

Shi offers a concise definition of voluntary simplicity as 'enlightened material restraint'. Variously defended by its advocates and practitioners on personal, communitarian, humanitarian, and ecological grounds, the Voluntary Simplicity Movement is based on the assumption that human beings can live meaningful, free, happy, and infinitely diverse lives, while consuming no more than an equitable share of nature. One study suggests that as many as 200 million people around the world are beginning their journey toward a 'simpler way' of life (Alexander and Ussher 2012).

The Voluntary Simplicity Movement (or something like it) will almost certainly need to expand, organise, radicalise, and politicise if anything resembling a post-growth or degrowth economy is to emerge through democratic processes. This highlights the political and macroeconomic significance of grassroots activity.

For more information:

Alexander, S. and Ussher, S. (2012) 'The voluntary simplicity movement: A multi-national survey analysis in theoretical context'. *Journal of Consumer Culture,* **12** (1): 66.

Shi, D. (2007). *The Simple Life: Plain Living and High Thinking in American Culture,* Athens: University of Georgia Press.

Alternatives to the consumer society

Is there an alternative to the consumer society, while still keeping a decent quality of life? In 1960 Cuba was blockaded by the United States (the 'Special Period') and exports dropped by 75%. It had to adapt to severe shortages of oil, medicine and food. However, Cuba now serves as an example of a country that has *thrived* despite having limited amounts of fossil fuels. Cuba has low per capita income, yet in quality of life it excels (Murphy and Morgan 2013). It is a materially poor country with 'First World' education, literacy and health care (Ibid.). It has maintained its human services programmes, free education, old age support, basic nutrition and free health care. The WWF Living Planet Report rated Cuba in 2006 as the only country to have genuine sustainable development (Ibid.). Cuba arguably represents an alternative, where material success (as measured by energy consumption) is secondary, while quality of life is given priority. Of course, I do understand that there remains the issue that Cuba is not a full democracy. However, Cuba's experience does show that humanity *can* survive (and even thrive) in a resource-constrained world (Ibid.). So it's not a case of having to go back and 'live in caves'. We can live a sustainable life with far less consumerism, less 'things', a much smaller ecological footprint – and still have a 'good life' (Washington 2015).

Solutions: dealing with the heresy of *more*

We need to deal with the 'heresy of more' (Washington 2015). 'More' is not always better, either for people or for stuff. It once used to be, but in today's world it actually makes us worse off. Yet our current civilisation seems locked in a trance of 'evermoreism' (Boyden 2004). 'More and more' people and possessions up to now has been seen as good. However, more and more things and junk means more consumption and greater use of resources. And that is good for the economy isn't it? Actually, in real terms it is not. Beyond a certain point, more people and more consumption brings *uneconomic* growth (Daly 2014). In real terms it is making society worse off, not better, and decreasing humanity's (and nature's) well-being (Ibid.). It is also markedly decreasing the prospects of future human well-being. We are stealing from posterity (Catton 2012). The big problems of overpopulation and overconsumption: 'have no technical fixes but only difficult moral solutions' (Daly 1991: 39).

Gowdy (2014: 40) invites us to do a thought experiment where (painlessly) humanity's population is returned to a few hundred million and ecosystems are restored. He notes however that if we keep the ideology of growth, accumulation and expansion, then we would be back to the current system in a few decades, too many people consuming too much and the Earth's life support systems teetering on collapse (Ibid.). Endless growth is the true heresy of 'more' that we must confront and change (Washington 2015).

Dealing with overpopulation will be hard, given the baggage that goes with it. Dealing with consumerism may be even harder, as it is actually easier to change family size than patterns of consumption (Campbell 2012). This may be partly because the advertising industry spends $500 billion a year urging us to consume more (Assadourian 2013). As the growth economy falters (as it is), we will increasingly be told that the 'only solution' is to consume more and more junk. However, it is a false solution, and the sooner we realise this, the better. In the past it took a decade or two to embed consumerism into the US psyche, which then got exported around the world. It will also take time to move away from rampant consumerism, to move back to 'thriftiness', to a more equitable level of consumption. We need to accept that we don't need to 'Shop till you drop!' or live like princes and princesses. Most (but not all) corporations will oppose change, as will the advertising industry, and probably also the car industry and the fast food industry. The list of those in opposition will undoubtedly stretch on and on. Sadly, 'time' is one of the things we don't have 'more' of, but the least of. To heal the world we need to act *now* to cure our consumer addiction (Washington 2015).

Most governments will sit on their hands and ignore the problems of over-population and overconsumption, while some will actively promote more population growth and more consumption in an attempt to increase their GDP and avoid a recession (Washington 2015). So, if we are going to break the addiction, change will have to *come from us*. We will have to demand a ban on 'planned obsolescence' of products (where they speedily wear out or break down). We will have to demand that society consumes less, and demand our governments reduce waste, and demand that corporate law be reformed. It won't be easy, but then the consequences of ignoring both overpopulation and overconsumption will be far worse. These are the key drivers of *un*sustainability (Ibid.). Some argue for action on one but not the other, but the two are conjoined twins in terms of environmental impact (see Chapter 4). We cannot accept the problems of one and ignore the other. We will not heal the world if we cannot recognise both as key problems, move past denial and solve these twin drivers of *un*sus-tainability. Without tackling these, any supposed 'sustainability strategies' to heal the world will remain tokenism.

Assumptions of hate

Society is full of 'assumptions of hate'. Such a claim may surprise some readers. However, most people understand that racism exists, where some people hate another section of our species. People will also have heard of misogyny (hatred of women), which can be part of patriarchy and even 'androcentrism' (placing a masculine point of view at the centre of one's world view, culture and history) (Curry 2011). Of course misandry (hatred of men) also exists, along with a corresponding 'gynocentrism' (a dominant or exclusive focus on women in theory or practice). There is also hatred of political groups (e.g. the Right, the Left, the Greens), hatred of religious groups and hatred of lifestyle choices (e.g. environmentalists, vegans, etc.). Finally, the strongest (and sadly most common) form of anthropocentrism is 'human supremacy' (Crist 2012), which amounts to a hatred (and fear) of nature (see Chapter 2). As you can see, humanity is very good at hating (and blaming) various human groups of the 'others', as well as the 'other' of the rest of life that Western society has cut itself off from (pretending to be the 'Masters', see Chapter 8). Hatred serves to alienate one from the hated group, so they cease to have the rights or values one might extend to other groups. They then become 'the other' that is deemed different from our social identity (Miller 2008).

I don't intend to spend a lot of space discussing these various hatreds. My point here is simply to remind the reader that assumptions of hate exist, and to point out that *all* these assumptions of hate hold us back from healing the world. Hatred creates polarisation, alienation, bigotry and denial, and holds us back from listening, dialogue and reinventing ourselves (see Chapter 7). Apart from anything else it is also self-corrosive psychologically and spiritually. Listening, dialogue and a 'duty of care' towards nature all involve a degree of *love* – as does healing our world.

Conclusion

Few of us stop to consider the assumptions that underlie what we think is true or 'how the world is'. Yet we should consider these, as Western society has developed a whole range of assumptions that are both false and unsustainable. Yet to question them will often get a response of outrage, as one is questioning the status quo and what many think of as 'common sense'. Of course the degree to which the world is wounded shows that it is well and truly time to question what we think is 'common sense'. If we are to heal the world we will need to transform our assumptions. We need an assumption that all human endeavours must operate within ecological limits. We need an assumption that nature has intrinsic value and deserves respect, and that we have a 'duty of care' towards her. We need an assumption that we seek 'enough' – not endless growth. We need an assumption that justice is something that applies to nature as well as humanity (and that the two must be entwined, Washington et al. 2018). Given the mass extinction event of life now underway (and caused by us), I believe we need a 'Nature First' assumption in regard to planning (Washington and Kopnina 2019). All these suggested assumptions are based on ecological reality, ecocentrism and ecological ethics, rather than the view that only humans matter.

What can *I* do?

In terms of the assumptions of neoclassical economics, there are many things we can do (see also Washington 2017):

1) Free yourself from the ideology of endless growth (given we live on a finite planet).
2) Support a move towards a 'steady state economy' (Daly 2014) to replace our endless growth economy (see www.steadystate.org).
3) Maintain that the market must be regulated to protect both society and nature. The market must reflect the truth in terms of social benefit and harm, along with the benefit or harm to nature.
4) Insist that environmental damage must no longer be considered a mere 'externality' to any economic model.
5) Explain that the idea we can substitute money or human capital for natural capital (= weak sustainability) is unsustainable and will just worsen the environmental crisis.
6) Point out that no economics should ignore physical laws and the reality of ecological limits (as neoclassical economics does).
7) Suggest that banks should be required to move (perhaps gradually) to a 100% reserve requirement, and make their money by financial intermediation and service charges, rather than lending at interest money they 'create out of nothing' (Daly 2000).
8) Urge society to regain control of the finance sector, remove its political power and ensure it serves a sustainable real world economy (Palley 2014).
9) Support increased overseas aid by the developed world for the developing world, but this must be specifically targeted to *ecologically sustainable transition projects* to an ecologically sustainable steady state economy.
10) Urge your Bureau of Statistics (or equivalent institution) to track and publish the 'Genuine Progress Indicator' (GPI) (Costanza et al. 2013). Society should also track the 'Happy Planet Index' (Layard 2005).
11) Support a tax on financial transactions, also known as a 'Tobin tax' where Tobin suggested it be 0.5% (Daly 2008). This will deter rapid speculative finance transfers that are part of 'extreme money' (Das 2011) and which exacerbate the debt crisis.
12) Support limits on income inequality such as the establishment of both minimum *and* maximum incomes in society (Washington 2017). Daly (2008) suggests a factor of 10–20 as the upper limit for maximum incomes over minimum. Czech (2013) suggest 15 times as the upper limit, while the

   Mondragon cooperative in Spain has a maximum pay of 9 times the minimum (Dietz and O'Neill 2013).

13) Support more flexible workdays and 'working from home' policies (Hein-
    berg 2011; Sukhdev 2013).

14) Support a sustainability tariff structure so that 'sustainable' countries are not
    disadvantaged (Heinberg 2011) when they trade with those that are not
    sustainable.

15) Speak in support of turning our economy from an ethics-free zone to one
    which is based on an Earth ethics (Curry 2011; Rolston 2012).

16) Support green investment (e.g. in renewable energy) and green jobs, which
    together can cushion any possible social impacts of moving to a steady state
    economy (Daly 2008).

In terms of consumerism we need to reinvent ourselves (see Chapter 7) so we:

17) Consume less, shop less and live more.

18) Argue against rampant consumerism and speak out in support of *thriftiness*.

19) Support reform of corporate law to support healing the world (Heinberg
    2011).

20) Argue for dematerialisation of the economy (Braungart and McDo-
    nough 2008). The developed countries should aim for a goal of Factor
    5 (use only 20% of current energy and resources, von Wiezsäcker et al.
    2009).

21) Concentrate on sufficiency and well-being (yours *and* nature's) rather than
    more 'stuff'.

22) Free yourself and friends from 'Shop till you drop!'

23) Free yourself from the dominance of the advertising industry. Don't watch
    advertising if you can avoid it, and seek to minimise the advertising you
    and your family are subjected to. Support the creation of an 'Advertising
    Tax' and billboard-free cities. Download the open-sourced software
    'AdBlock Plus' for your Internet Browser. This free add-on will remove all
    advertising from websites (including videos).

24) Rethink, reduce, reuse and recycle in regard to the products in your living
    place.

25) Argue in favour of 'choice editing' so that unsustainable products are banned.

26) Argue in favour of 'Extended Producer Responsibility' to control the
    planned obsolescence common in business today (Washington 2015).

27) Speak out to reduce (not expand) resource use, both non-renewable and
    renewable. For non-renewable resources a 'depletion quota' has been suggested
    (Daly 1991) or a 'severance tax' at the mine-mouth or well-head (Daly 2008).

In terms of ideologies of hate:

28) Stop to think about what assumptions you may have been taught that you
    really should question.

29) Ask yourself if there are any groups you 'hate'? Is this a useful or healing approach?
30) Consider how you can help to break down polarisation caused by hatred, and instead create dialogue and healing.

References

Assadourian, E. (2010) 'The rise and fall of consumer cultures', in *2010 State of the World: Transforming Cultures from Consumerism to Sustainability*, eds. L. Starke and L. Mastny, London: Earthscan, pp. 3–20.

Assadourian, E. (2013) 'Re-engineering cultures to create a sustainable civilization', in *State of the World 2013: Is Sustainability Still Possible?*, ed. L. Starke, Washington: Island Press, pp. 113–125.

Boyden, S. (2004) *The Biology of Civilisation: Understanding Human Culture as a Force in Nature*, Sydney: UNSW Press.

Braungart, M. and McDonough, W. (2008) *Cradle to Cradle: Remaking the Way We Make Things*, London: Vintage Books.

Campbell, M. (2012) 'Why the silence on population?', in *Life on the Brink: Environmentalists Confront Overpopulation*, eds. P. Cafaro and E. Crist, Georgia, USA: University of Georgia Press, pp. 41–55.

Cardinale, B. J., Duffy, E., Gonzalez, A. et al. (2012) 'Biodiversity loss and its impact on humanity'. *Nature*, **486** (7401): 59–67. DOI: 10.1038/nature11148.

Catton, W. (2012) 'Destructive momentum: Could an enlightened environmental movement overcome it?', in *Life on the Brink: Environmentalists Confront Overpopulation*, eds. P. Cafaro and E. Crist, Georgia, USA: University of Georgia Press, pp. 16–28.

Collins, P. (2010) *Judgment Day: The Struggle for Life on Earth*, Sydney: UNSW Press.

Common, M. and Stagl, S. (2005) *Ecological Economics: An Introduction*, Cambridge: Cambridge University Press.

Costanza, R., Alperovitz, G., Daly, H. et al. (2013) 'Building a sustainable and desirable economy-in-society-in-nature', in *State of the World 2013: Is Sustainability Still Possible?*, ed. L. Starke, Washington: Island Press, pp. 126–142.

Crist, E. (2012) 'Abundant Earth and the population question', in *Life on the Brink: Environmentalists Confront Overpopulation*, eds. P. Cafaro and E. Crist, Georgia: University of Georgia Press, pp. 141–151.

Curry, P. (2011) *Ecological Ethics: An Introduction*, Second edition. Cambridge: Polity Press.

Czech, B. (2013) *Supply Shock*. Gabriola, BC: New Society Publishers.

Daly, H. (1991) *Steady State Economics*, Washington: Island Press.

Daly, H. (1996) *Beyond Growth: The Economics of Sustainable Development*, Boston: Beacon Press.

Daly, H. (2008) 'A steady-state economy: A failed growth economy and a steady-state economy are not the same thing; they are the very different alternatives we face', 'think-piece' for the Sustainable Development Commission, UK, April, 24, 2008, see: http://steadystaterevolution.org/files/pdf/Daly_UK_Paper.pdf (accessed 27 March 2019).

Daly, H. (2014) *From Uneconomic Growth to the Steady State Economy*, Cheltenham: Edward Elgar.

Daly, H. and Cobb, J. (1994) *For the Common Good: Redirecting the Economy toward Community, the Environment, and a Sustainable Future*, Boston: Beacon Press.

Das, S. (2011) *Extreme Money: Masters of the Universe and the Cult of Risk*, Oakland, NJ: FT Press.

Diesendorf, M. and Hamilton, C. (1997, eds.) *Human Ecology, Human Economy*, Sydney: Allen & Unwin.

Dietz, R. and O'Neill, D. (2013) *Enough Is Enough: Building a Sustainable Economy Is a World of Finite Resources*, San Francisco: Berrett-Koehler Publishers.

Ekins, P. (1991) 'The sustainable consumer society: A contradiction in terms?', *International Environmental Affairs*, **3**: 243–257.

Ellul, J. (1975) *The New Demons*, New York: Seabury Press.

Flavin, C. (2010) 'Preface', in *State of the World 2010: Transforming Cultures from Consumerism to Sustainability*, eds. L. Starke and L. Mastny, New York: Worldwatch Institute/ Earthscan, pp. xvii–xx.

Foxon, T., Köhler, J., Michie, J. and Oughton, C. (2012) 'Towards a new complexity economics for sustainability', *Cambridge Journal of Economics*, **37**: 187–208.

Georgescu-Roegen, N. (1971) *The Entropy Law and the Economic Process*, Cambridge, MA: Harvard University Press.

GFN. (2019) 'Country trends', Global Footprint Network, see: http://data.footprintnet work.org/#/countryTrends?cn=5001&type=BCpc,EFCpc (accessed 27 March 2019).

Gintis, H. (2011) 'Gene-culture coevolution and the nature of human sociality'. *Philosophical transactions of the Royal Society of London. Series B, Biological Sciences*, **366**: 878–888.

Gowdy, J. (2014) 'Governance, sustainability, and evolution', in *State of the World 2014: Governing for Sustainability*, ed. L. Mastny, Washington: Island Press, pp. 31–40.

Graff, J. (2010) 'Reducing work time as a path to sustainability', in *State of the World 2010: Transforming Cultures from Consumerism to Sustainability*, eds. L. Starke and L. Mastny, New York: Worldwatch Institute/Earthscan, pp. 173–177.

Hamilton, C. (2010) *Requiem for a Species: Why We Resist the Truth about Climate Change*, Australia: Allen and Unwin.

Hardin, G. (1968) 'The tragedy of the commons', *Science*, **162** (859): 1243–1248.

Heinberg, R. (2011) *The End of Growth: Adapting to Our New Economic Reality*, Canada: New Society Publishers.

Higgs, K. (2014) *Collision Course: Endless Growth on a Finite Planet*, Cambridge, MA: MIT Press.

Kopnina, H. and Blewitt, J. (2015) *Sustainable Business: Key Issues*, London: Routledge.

Layard, R. (2005) *Happiness: Lessons from a New Science*, New York: Penguin Press.

Lebow, V. (1955) 'Price competition in 1955', *Journal of Retailing*, Spring, 1955, see: www. gcafh.org/edlab/Lebow.pdf (accessed 5 February 2019).

Miller, J. (2008). 'Otherness', in *The SAGE Encyclopedia of Qualitative Research Methods*, Thousand Oaks: SAGE Publications, Inc, pp. 588–591.

Moore, J. and Rees, W. (2013) 'Getting to one-planet living', in *State of the World 2013: Is Sustainability Still Possible?*, ed. L. Starke, Washington: Island Press, pp. 39–50.

Murphy, P. and Morgan, F. (2013) 'Cuba: Lessons from a forced decline', in *State of the World 2013: Is Sustainability Still Possible?*, ed. L. Starke, Washington: Island Press, pp. 332–341.

Oelschlaeger, M. (1991) *The Idea of Wilderness: From Prehistory to the Age of Ecology*, New Haven/London: Yale University Press.

Oreskes, N. and Conway, M. (2010) *Merchants of Doubt: How a Handful of Scientists Obscured the Truth on Issues from Tobacco Smoke to Global Warming*, New York: Bloomsbury Press.

Ostrom, E. (1990) *Governing the Commons: The Evolution of Institutions for Collective Action*, Cambridge: Cambridge University Press.

Palley, T. (2014) 'Making finance serve the real economy', in *State of the World 2014: Governing for Sustainability*, ed. L. Mastny, Washington: Island Press, pp. 174–180.

Prugh, T. (2013) 'Getting to true sustainability', in *State of the World 2013: Is Sustainability Still Possible?*, ed. L. Starke, Washington: Island Press, pp. 111–112.

Rees, W. (2008) 'Toward sustainability with justice: Are human nature and history on side?', in *Sustaining Life on Earth: Environmental and Human Health through Global Governance*, ed. C. Soskolne, New York: Lexington Books, pp. 81–94.

Rees, W. (2010) 'What's blocking sustainability? Human nature, cognition and denial', *Sustainability: Science, Practice and Policy*, **6** (2) (ejournal), see: http://sspp.proquest.com/archives/vol6iss2/1001-012.rees.html (accessed 21 March 2019)

Rees, W. (2019) 'End game: The economy as eco-catastrophe and what needs to change', *Real World Economic Review*, **87**: 132–148.

Renner, M. and Prugh, T. (2014) 'Failing governance, unsustainable planet', in *State of the World 2014: Governing for Sustainability*, ed. L. Mastny, Washington: Island Press, pp. 3–19.

Rolston, H., III. (2012) *A New Environmental Ethics: The Next Millennium of Life on Earth*, London: Routledge.

Starke, L. and Mastny, L. (2010) *State of the World 2010: Transforming Cultures from Consumerism to Sustainability*, New York: Worldwatch Institute/Earthscan.

Stiglitz, J. E. (2002) *Globalization and Its Discontents*, New York: Allen Lane.

Sukhdev, P. (2010) 'Preface', in *The Economics of Ecosystems and Biodiversity: Ecological and Economic Foundations*, ed. P. Kumar, London: Earthscani, pp. xvii–xxvii.

Sukhdev, P. (2013) 'Transforming the corporation into a driver of sustainability', in *State of the World 2013: Is Sustainability Still Possible?*, ed. L. Starke, Washington: Island Press, pp. 143–153.

Tacey, D. (2000) *Re-Enchantment: The New Australian Spirituality*, Australia: Harper Collins.

Victor, P. (2008) *Managing without Growth: Slower by Design, Not Disaster*, Cheltenham, UK: Edward Elgar.

Victor, P. (2019) *Managing without Growth: Slower by Design, Not Disaster*, Second edition. Cheltenham, UK: Edward Elgar.

von Weizsäcker, E., Hargroves, K., Smith, M., Desha, C. and Stasinopoulos, P. (2009) *Factor 5: Transforming the Global Economy through 80% Increase in Resource Productivity*, UK: Earthscan.

Washington, H. (2013) *Human Dependence on Nature: How to Help Solve the Environmental Crisis*, London: Earthscan.

Washington, H. (2015) *Demystifying Sustainability: Towards Real Solutions*. London: Routledge.

Washington, H. (2017, ed.) *Positive Steps to a Steady State Economy*, Sydney: CASSE NSW, see: https://steadystatensw.files.wordpress.com/2017/06/posstepsroyal11ptjustheaderfinaljune12thebooklowres.pdf.

Washington, H. (2018) *Kinship Poems*. Lulu.com, see: www.lulu.com/shop/haydnwashington/kinship-poems/paperback/product-23580118.html   (accessed 27 March 2019).

Washington, H. (2019) Ecosystem services – A key step forward *or* anthropocentrism's "Trojan Horse" in conservation?', in *Conservation: Integrating Social and Ecological Justice*, eds. H. Kopnina and H. Washington, New York: Springer, pp. 73–90.

Washington, H., Chapron, G., Kopnina, H., Curry, P., Gray, J. and Piccolo, J. (2018) 'Foregrounding ecojustice in conservation', *Biological Conservation*, **228**: 367–374.

Washington, H. and Kopnina, H. (2019) 'Conclusion: A just world for life?', in *Conservation: Integrating Social and Ecological Justice*, eds. H. Kopnina and H. Washington, New York: Springer, pp. 219–228.

Washington, H. and Twomey, P. (2016, eds.) *A Future beyond Growth: Towards A Steady State Economy*, London: Routledge.

Westra, R. (2008) 'Market society and ecological integrity: Theory and practice', in *Sustaining Life on Earth: Environmental and Human Health through Global Governance*, ed. C. Soskolne, New York: Lexington Books, pp. 41–50.

Whitehead, A. (1929) *Process and Reality*, New York: Harper Brothers.

Wijkman, A. and Rockstrom, J. (2012) *Bankrupting Nature: Denying Our Planetary Boundaries*, London: Routledge.

Wilkinson, R. and Pickett, K. (2010) *The Spirit Level: Why Equality Is Better for Everyone*, London: Penguin Books.

World Bank. (n.d.) 'Gini coefficient (World Bank Estimate)', see: https://data.worldbank.org/indicator/SI.POV.GINI?end=2015&locations=TH-CN-AU-US-JP-BT-ZA-GB&start=1981&view=chart (accessed 15 March 2019).

6

APPROPRIATE TECHNOLOGY TEMPERED BY HUMILITY

I consider our predicament …
Where climate now is
Close to the
Runaway point;
Where half of life
May be sent extinct
By century's end;
And what of
Chernobyl, Fukushima,
DDT, PCBs?
And micro-plastics
Invading our very cells?
Maybe technology
Needs ethics
To guide it?
Maybe tech cannot
Solve everything?
Perhaps it must be
Appropriate
To serve society?
To serve life?

From 'Appropriate' (Washington 2019)

SUMMARY

Many people in Western society think all technology is good. However, technology can be either appropriate or inappropriate in terms of healing the world. Key problems are the ideologies of techno-centrism (and techno-optimism) and ecomodernism, which assume that technology will solve all our problems. Our worsening world predicament however shows that this is mistaken. Ecomodernism is discussed in detail, showing that the 'Ecomodernist Manifesto' is in fact based on denial of the environmental crisis and on ideological (not factual or eco-ethical) premises. Renewable energy is shown to be a very appropriate technology and probably our best solution to solve the climate crisis. However, there are also several very *in*appropriate technologies being promoted by sections of society, the key one of which is nuclear power. Compared to renewable energy, nuclear is more expensive, slower to install and much more dangerous. It is never a good idea to replace one dangerous technology (use of fossil fuels) with another dangerous technology (nuclear power). For the detail explaining this summary, please read on; however, if you just want to know the things *you can do* (five are listed) about this, then turn to the end of the chapter.

Introduction

A mindless commitment to any or all 'technology' holds us back from healing the world. As we shall see, it encourages us to believe we are 'Masters of the Universe' and in control of everything. In fact, such an ideology just exacerbates a dangerous mindset and distracts us from more important things – such as healing the world. I should point out that I am not a Luddite, and this chapter is not claiming all technology is bad (or that humanity should go back to living in caves). However, we do need to closely examine the 'cult' of technology to understand its limits and weaknesses. Most importantly, we need to look at the many technologies in existence and ask for each one 'Is this appropriate?' Appropriate to what? Appropriate to the predicament humanity finds itself in (Washington 2015). It is time now for technologies that help us to heal the world, not push society, and the nature that supports it, closer to collapse.

What is appropriate?

We need to discuss the part technology plays in healing or harming the world. Techno-centrists see technology as a god that will solve all problems. However, former President Obama's Chief Science Advisor, John Holdren, noted in 2007 that: 'belief in technological miracles is generally a mistake' (in Nicholson 2013: 317). The reason is because technology alone is not a meaningful basis for deciding if something is good or healing (Washington 2015). Meadows et al. (2004) note that technological advance, or markets by themselves, unchanged, unguided by understanding, respect or commitment to sustainability, *cannot* create a sustainable society. Others again see technology as the cause of the environmental crisis, that it is inherently flawed, being part of a 'scientism'[1] (Daly 1991; Curry 2011) that is destroying the Earth. Who is right?

'Technology' is generally lumped together as just one thing, when in fact it is not uniform and consistent. Some parts can be key solutions that help us heal the environmental crisis. Others are entirely the opposite. Technology thus has promising aspects and very worrying ones. Chapter 8 shows that the 'Mastery of Nature' aspect of some science and technology has led us close to disaster. But this is an 'ideology' pasted onto science, not necessarily something inherent in either science or technology (Washington 2015). Science and technology do not have to be this way. Part of the problem has not been science or even technology per se, but that they have not been *appropriate*. Appropriate to the reality we face: escalating environmental and climate crises. Appropriate to the rapidly reducing time-frame over which we need solutions. Appropriate for a healing worldview and an 'Earth ethics' (Rolston III 2012). Schumacher (1973) rightly called for 'appropriate' technology to lead us to sustainability. We need an environmental revolution, but unlike the Industrial Revolution, this one must use appropriate technology. However, it should also be driven by our need to make peace with nature (Brown 2006; Washington 2018).

One of the reviewers of this book reminded me that it is important to note here that there are degrees as to what is *truly* 'appropriate' technology. For example, one could dry clothes using an electric dryer powered by solar electricity; however, it would be even more appropriate to use a clothesline. One could buy an electric car that one charges from solar panels, but of course using a bicycle would be even more appropriate for many needs. One could supply energy to a normal energy-wasting house by extensive solar panels – or one could design one's house to *minimise energy use* and use far less panels. We have become so used in the West to using labour-saving devices, when now we should be concentrating on lifestyles that minimise both material and energy use. There are many ideas out there that assist, such as using biogas (e.g. Alexander 2018). Websites such as 'Resilience' (www.resilience.org/act-resources/) offer suggestions on how to be more truly appropriate in terms of what we do.

So in terms of appropriate technology – are there solutions? Yes, there most certainly are. Washington (2015) argues this is one of the most frustrating aspects of the whole debate, the denial of real and workable solutions. The environmental crisis is not an irrevocable 'Decree of Fate' to which we must submit. It is a human–caused problem and has human–invented solutions, *provided we act* (Ibid.). One of the most frustrating aspects is that denial of appropriate technology (such as renewable energy) ignores the fact that this will create a large number of green jobs, more jobs than the fossil fuel and mining industries combined (Ibid.). Yet many conservatives who supposedly support 'job creation' will (in the same breath) oppose renewable energy.

Techno-centrism and ecomodernism

Humanity has come a long way from using stone tools. We have learned a lot of science, but along with this we have developed a powerful technology that has transformed the world – but overall not for the better (Washington 2015). Why do I say it has not been for the better? After all, it is commonly argued that technology has allowed society to do wonderful things and expand to many places humans never lived before. It has solved many medical problems too. The question is whether that transformation has been uniformly positive – or whether it has created more bads than goods? The point I wish to make here is that techno-centrism and techno-optimism are dominant ideologies in our society (Mason 2012). The assumption is that 'technology is good', and even more so that technology will solve all our problems. A related term to techno-centrism is Cornucopianism. A Cornucopian is a futurist who believes that continued progress and material wealth will be met by continued advances in technology. Fundamentally, they believe that there is enough matter and energy on the Earth to provide for an ever-rising population and consumption. The term comes from the *Cornucopia*, the 'horn of plenty' of Greek mythology, which magically supplied its owners with endless food and drink. The Cornucopians do not see resource depletion as limiting human progress, and see technology as 'saving the day' through greater development and substitution of resources. Techno-centrism and Cornucopianism remain dominant assumptions in Western society, yet as Chapter 5 explains, this is one of the unsustainable assumptions that bedevils society. The assumption that technology is always positive is clearly not just questionable – but mistaken.

'The Ecomodernist Manifesto' (EM 2015) was written by several academics who came from an anthropocentric and techno-centric worldview. It argues (p. 9):

> Despite frequent assertions starting in the 1970s of fundamental 'limits to growth', there is still remarkably little evidence that human population and economic expansion will outstrip the capacity to grow food or procure critical material resources in the foreseeable future.

Despite the mute testimony of *all* the environmental indicators shown in Chapter 1, ecomodernism is thus in outright denial of ecological and biophysical limits. Ecomodernism is also strongly anthropocentric and describes the world as 'the human planet'. The Ecomodernist Manifesto (EM 2015: 7) argues:

> Intensifying many human activities – particularly farming, energy extraction, forestry, and settlement – so that they use less land and interfere less with the natural world is the key to decoupling human development from environmental impacts. These socioeconomic and technological processes are central to economic modernization and environmental protection. Together they allow people to mitigate climate change, to spare nature, and to alleviate global poverty.

There seems to be a strong denial in operation in this Manifesto, which is ideologically (not factually or eco-ethically) driven. Clearly, the proponents of ecomodernism are strongly anthropocentric, techno-centric and techno-optimistic. They don't believe in environmental science or accept what environmental indicators are telling society. They also don't seek to control overpopulation and overconsumption – as they support growthism (which is what 'intensification' is all about). They also clearly don't believe we need to transform the unsustainable endless growth economy (see Chapters 3 and 5). They are also clearly very pro-nuclear power (BI n.d.) seeking a huge expansion of this. Many inaccuracies or illogicalities occur in their 'intensification' rhetoric, to wit:

*   **Intensive agriculture** – this means greater use of fertilisers and pesticides (neither of which are sustainable). Intensive fertiliser use is already responsible for humanity exceeding two planetary boundaries, being nitrogen and phosphorus pollution (Steffen et al. 2015). Intensive agriculture will cause even greater soil loss than the current 75 billion tonnes/yr (Brown 2011). Some argue that soil erosion is so bad we may have only 60 years of farming left on arable soils (Arsenault 2017). It will mean more water use and increase the decline of aquifers when already half the world's people live in countries where water tables are falling as aquifers are depleted (Brown 2011). It will mean more salinisation and soil acidification (almost certainly due to the increasing use of nitrate fertilisers, Guo et al. 2010). It would probably mean massive hydroponics installations with plastic greenhouses, some of which will seek to source their water from desalinisation plants (requiring a great deal of energy). It represents the opposite of organic agriculture and an ecologically sustainable (or regenerative) approach.
*   **Intensive energy extraction** – this phrase in fact represents a deliberate mystification of ecomodernists' strong pro-nuclear stance. Indeed the Manifesto (EM 2015: 23) inaccurately states: 'Nuclear fission today represents the only present-day zero-carbon technology with the demonstrated ability to meet most, if not all, of the energy demands of a modern economy'. These claims are shown to be incorrect later in the 'inappropriate technologies' section. Ecomodernists unashamedly argue for a 'high energy planet' (Trembath 2013) – even though throughout human history, greater energy use has *always* gone hand in hand with greater environmental impact (Boyden 2004). Yet ecomodernists seek to convince us that in a 'high energy planet', this greater energy use will somehow not be used to accelerate the human expansion phase even further. In fact, a high energy planet would operate by clearing more land and building more infrastructure such as roads that already represent a major threat to biodiversity (Laurance et al. 2018; Alamgir et al. 2019). Human history to date clearly shows that greater energy use equals greater destruction of nature. If ecomodernists actually believe the Earth is 'the human planet' then greater energy use will continue the domestication (= degradation) of the last wild areas left on Earth. What we need is *less energy use* by society as it abandons unsustainably high energy lifestyles (Washington 2015).
*   **Intensive forestry** – At least 17.7 million hectares of forest are cleared each year (WWF n.d.) or 47 football fields a minute. However, the

Ecomodernist Manifesto (EM 2015) fails to call for what is actually needed, which is intensive *reforestation* of land. Sadly, deforestation makes money for governments and hence is open to being overlooked (Thomasberger 2019); at worst it is open to corruption. Intensive forestry is likely just to accelerate deforestation, not heal through large-scale reforestation with native species.

- **Intensive cities** – ecomodernists don't seek to control overpopulation and yet call for more and more people to move into cities (EM 2015), ostensibly arguing that natural areas will have less pressure put upon them (Meyer 2013). However, cities have huge ecological footprints that still put huge pressure on natural areas. For example, the ecological footprint of Athens is 122% that of the whole of Greece, and Cairo's is 84% of that of all Egypt (GFN 2015). The ecological footprint of London is 120 times its area, and that of Tokyo is three times the area of *all of Japan* (GDRC n.d.). Because the impact is happening out of sight of the city does not mean the impact is any less. While there are ways of making cities more sustainable (WWI 2016), the fact remains that it is dubious whether large cities can ever be seen as properly ecologically sustainable in the long term. Cities require cement, sand, aggregate, ores for steel and copper and other metals, timber, food and many other resources, all sourced from wild nature and agroecosystems. Cities also act to strongly reduce human interaction with nature, and such interaction is very much needed to avoid 'Nature Deficit Disorder' (Louv 2005). It is encouraging to hear recently that some academics seek to promote 'post-anthropocentric cities' (Yigitcanlar et al. 2018). However, what the world actually needs is rapid population stabilisation, then a downward shift over the next century towards an ecologically sustainable population. If we stabilise world population at 8 or 9 billion we could reduce it to 6 billion by the end of the century and to a sustainable 2–3 billion by the end of the following century (Staples and Cafaro 2012: 296). The 'intensive cities' idea is clearly part of ecomodernists' 'high energy planet' agenda, aimed at supporting massive nuclear power plant construction. The many problems of nuclear power are discussed later.

It is thus not true that these four key strategies of ecomodernism will 'interfere less with the natural world' – despite the rhetoric of the Manifesto, the reality is quite the opposite. It is a public relations approach that seeks to convince readers through confident-sounding but misleading words. In fact it is part of the 'non-denier denier' approach (Hoggan 2009) where proponents *say* they accept the reality of environmental problems but then fail to support any practical solutions (or suggest such solutions would be too expensive). The reason I believe ecomodernists take such a stance is because they are addicted to the idea of endless growth. Rather than being 'central' to environment protection (as ecomodernists claim), their strategies are ecologically *un*sustainable and will accelerate ecocide and extinction. The Breakthrough Institute is a key spokesperson for ecomodernism, and recently ran a symposium 'Ecomodernism 2018: Achieving

Disagreement' (BI 2018) that argued the idea that intensive agriculture and intensive urbanisation (and energy use) would allow society to support the 'Half Earth' (Wilson 2016) vision where we reserve half of all lands for nature. Clearly, some who attended were sincere in their support for the 'Half Earth' vision. However, to summarise my points above, the strategies of ecomodernists such as the Breakthrough Institute will actually fail to allow such a vision to happen as: 1) they don't support control of overpopulation; 2) they don't support control of overconsumption and growth economics; and 3) they remain wedded to a high energy, high technology, high consumption vision. This will only increase ecocide and make it impossible to reach a 'Half Earth' vision.

Ecomodernists claim they are 'eco-pragmatists' who: 'describe our vision for putting humankind's extraordinary powers in the service of creating a good Anthropocene' (EM 2015: 7). They thus imply that they 'do good'. In contradiction to this, I believe their techno-centric stance and denial of ecological reality will just accelerate ecocide and hinder the healing of our world.

BOX 6.1 SUSTAINABILITY PARADOXES, HELEN KOPNINA

Sustainability bottlenecks, paradoxes or myths are found in the broad field of sustainability. Yet, few authors come up with a clear framework for sustainable solutions. The concept of sustainability – the way most of us use it today – emerged in the 1960s in response to concern about environmental degradation and social equity. Yet, sustainability is not always perceived in a progressive context. Sometimes, overwhelmed by the challenges of *un*sustainability, some of us have despaired at the possibility of addressing this – either because we feel individually powerless to change anything, or because of the lack of interest. Others in the community feel that sustainability challenges, such as overpopulation and overconsumption, are grossly exaggerated. And there is a group – to which particularly large corporations and their public relations belong – that have expressed their optimism about their own contribution to sustainability, assuming that no further effort is necessary.

In regard to eco-efficiency, the 'Cradle to Cradle' and 'Circular Economy' concepts have been suggested. Based on the notion of *industrial metabolisms*, William McDonough and Michael Braungart developed the notion of the Cradle to Cradle (C2C). They ask us to contemplate not just minimizing the damage, the way eco-efficiency does, but eliminating it all together. In this view, eco-efficiency simply serves to 'slow the process of destruction' and 'makes a bad design last longer'. Instead, the C2C framework propagates a simple dictum: Waste equals Food. A cherry tree metaphor exemplifies the example of the C2C principle, with its 'waste' (blossoms and berries) either directly consumed by birds, or decomposing into food for soil, with nutrients flowing indefinitely in cycles of birth, decay and rebirth. Understanding these regenerative systems can allow engineers and designers to recognize that all materials can be designed as nutrients that flow through natural (biological) or designed (technological) metabolisms.

One of the contradictions in sustainability involves the desire to: 1) improve human health (resulting in population growth) and material well-being (resulting in increase in consumption); and 2) secure the carrying capacity of this planet (which is conditional on halting population and consumption). This is something of a 'have your cake and eat it' oxymoronic objective. While many policy-makers and members of civil society are aware of the symptoms of unsustainability, few of them have realized the core challenges of unsustainability. Few sustainability programs have addressed the root causes of poverty, part of which is population growth itself, and the global spread of unsustainable practices. While many have accepted poverty elevation and economic aid to developing countries as normative, few invest in family planning or attempt to halt production of 'cradle to grave' products. In order to achieve sustainability aims, public, corporate and government stakeholders need to learn to better understand not only challenges, but also the mechanisms underlying unsustainable practices. Once these challenges and mechanisms are understood, a more positivistic turn toward solutions becomes possible. This would help to heal the environmental crisis.

For more information: https://www.leidenanthropologyblog.nl/articles/para doxes-of-sustainability

Renewable energy: an appropriate technology

As of 2016, renewable energy accounted for an estimated 18.2% of global total final energy consumption, with modern renewables representing 10.4% (REN21 2018). Renewable energy made up 70% of net additions to global power capacity in 2017 (Ibid.). IRENA (2018) notes:

> Electricity from renewables will soon be consistently cheaper than from fossil fuels. By 2020, all the power generation technologies that are now in commercial use will fall within the fossil fuel-fired cost range, with most at the lower end or even undercutting fossil fuels.

Jacobson et al. (2018) note that in normal traditional economic terms, renewables are now a similar price to the current 'business-as-usual' fossil fuel systems. However, they point out that when one considers the *full costs* of energy + health + climate, renewables are only a *quarter* the cost of the current fossil fuel systems. They conclude (p. 247) that it is fully possible to reach: 'a fully integrated all-sector 100% clean, renewable, efficient, and reliable energy infrastructure by 2050, if not sooner'.

The main types of renewables are summarised below:

1   **Windpower**. The cost of wind energy has come down hugely since the late 1990s. IRENA (2018) lists the cost of wind power as 6 c/kWh or less in 2017 (cheaper than both fossil fuel or nuclear power). Globally there is now 539 GW of windpower installed as of 2017 (REN21 2018).

2   **Solar**. Solar thermal is ideal for water heating and can make a significant contribution to space heating and cooling. Electricity from the sun can be produced either by concentrated solar thermal (CST) or by solar photovoltaic (PV). At the end of 2017 there were 402 GW of solar PV, and 4.9 GW of solar CST (REN21 2018). However, it is worth noting there was also 472 GW thermal of hot water capacity (Ibid.). The two main past problems of solar – remoteness from markets and the need for flexible back-up or storage to counter its intermittent nature – have now been largely solved (Diesendorf 2014). The cost of solar PV in 2017 had fallen to 10 c/kWh (IRENA 2018). This is cheaper than both fossil fuel and nuclear power.

3   **Geothermal and hot rocks**. Geothermal traditionally harnesses the steam produced naturally by the Earth's heat. Global generating capacity in 2017 was 12.8 GW (REN21 2018), almost all being conventional geothermal, where hot water reaches the Earth's surface in volcanic regions. However, there is also great potential for 'hot rock' power systems (*The Guardian* 2018).

4   **Hydroelectricity**. Hydroelectricity is currently the most developed form of renewable electricity, producing 1,114 GW in 2017 (REN21 2018). This still has large potential in Africa and parts of Asia. However, large hydroelectricity dams (such as the Three Gorges in China and Sardar Sarova in India) generally have major adverse environmental and social impacts (UCS n.d.). Small- and medium-scale hydro projects in widely distributed areas are likely to be better options, though these too can have issues (e.g. siltation and eutrophication) that need to be considered (Makhijani and Ochs 2013).

5   **Wave and tidal power (and ocean currents)**. The power of the tides has been harnessed in a few places in the world (e.g. the Rance Estuary in France) with around 0.5 GW in operation in 2017 (REN21 2018). However, conventional tidal power is limited geographically to regions where very high tides prevail. Waves are found on all coastal regions, but wave power has been harder to harness. However, there are now many promising projects underway around the world.

6   **Bioenergy**. Globally, photosynthesis stores solar energy in biomass (organic material) at a rate of about seven times the 2010 rate of global energy use (GEA 2012). Thus there is a substantial biomass resource worldwide. Plants store solar energy in wood, fibre and oils. These can be burned to produce energy for both stationary sources and transport, or converted to liquid and gaseous fuels by a wide range of processes. There was 122 GW of bio-power capacity in 2017 (REN21 2018). To be sustainable, however, bioenergy has to be properly managed (Diesendorf 2007; WWF 2011).

Inappropriate technologies

There are several inappropriate technologies to be considered, especially those based on the anthropocentric ideology of control of nature. *Geoengineering* is one, as it would be likely to have massive side-effects (Washington 2015). It would also entrench the use of fossil fuels and exacerbate the climate crisis (CIEL 2019). *Carbon capture and storage* is another inappropriate technology as: 1) it is difficult to capture all the CO_2 in power station emissions; 2) it requires a huge amount of energy to liquefy and pump the CO_2 into the ground; 3) there will almost certainly be significant leakage; and 4) it requires special geology not found everywhere. This is discussed in more detail in Pittock (2009) and Baxter (2017). However, the key inappropriate technology I consider here is *nuclear power*, given that many lobby groups (such as ecomodernists) continue to push for its huge expansion. The key point to remember is that renewable energy is both cheaper, safer and far more ethically sound than nuclear. Nuclear power is an *in*appropriate technology. It is not the answer because you don't solve one serious problem by creating another very serious problem. Energy expert Lovins (2006: 6) has rightly asked about nuclear power:

> Why keep on distorting markets and biasing choices to divert scarce resources from the winners to the loser – a far slower, costlier, harder, and riskier niche product – and paying a premium to incur its many problems?

So why is society debating its merits? The reason is the powerful nuclear and uranium-mining lobbies (and ecomodernists), which (due to concern over climate change) see a chance to resurrect their nuclear fantasy. Washington (2015) summarises the drawbacks of nuclear energy:

- It's *not carbon neutral anyway*, and can produce 10–40% as much CO_2 as a coal-fired power station, depending on the uranium ore grade (Mudd and Diesendorf 2008). Nuclear energy results in 9–25 times (Jacobson and Delucchi 2011) or 6–22 times (Diesendorf 2014) more carbon emissions than wind energy, depending on the grade of the uranium ore. With low grade ore, the CO_2 emissions of nuclear power stations are comparable with those from a natural gas power station. Hence nuclear energy, based on existing commercial technology, cannot be a long-term energy/electricity solution to global climate change (Diesendorf 2014).
- Nuclear energy is *too slow to deploy*. It takes many years to commission a nuclear power station. The historic 'planning-to-operation' time for new nuclear plants has been 11–19 years, compared with an average of 2–5 years for wind and solar installations (Jacobson and Delucchi 2011).
- Nuclear power is *not cheap*. The current cost of nuclear is estimated to be 17–34 c/kWh (Diesendorf 2014). In 2017, wind power cost had dropped to 6 c/kWh and solar PV was 10 c/kWh (REN21 2018). Nuclear is thus

much more expensive than wind power and large-scale solar. However, the cost of solar is dropping all the time (IRENA 2018), while nuclear costs are likely to keep rising (see the point below about decommissioning reactors). If the huge subsidies given to nuclear power are taken into account (Diesendorf 2013) and the hidden external costs (Sovacool 2011), nuclear is clearly even *more* expensive than solar electricity.

- Conventional nuclear fission relies on *finite reserves of high-grade uranium*. A large-scale conventional nuclear programme would exhaust these in less than a century (Jacobson and Delucchi 2011). There are not enough total uranium reserves to provide a low-carbon future anyway, unless we move to 'fast breeder' reactors. These have major safety problems with the potential for a large (non-nuclear) explosion that could disperse radioactive material into the environment (Kumar and Ramana 2009). By creating much larger quantities of plutonium than conventional reactors, they also increase the risk of nuclear proliferation (Jacobson and Delucchi 2011). The world's largest fast breeder reactor, the French 'Superphoenix', was launched in 1974, connected to the grid in 1986, and closed in 1998 after many technical problems (Diesendorf 2014).

- *Proliferation of nuclear weapons* is a major problem. Even a regional nuclear war (such as one between India and Pakistan) could bring on a 'nuclear winter' resulting in global agricultural collapse for several years and mass starvation (Robock and Toon 2010). There is ample evidence that so-called 'peaceful' nuclear power has contributed, and still contributes, to the proliferation of nuclear weapons (Diesendorf 2014). It thus increases the risk of a future nuclear war.

- Nuclear power stations produce *highly radioactive waste* that must be separated from the environment for at least a 100,000 years.

- Nuclear power stations cannot blow up like an atomic bomb, but *can* melt down like Chernobyl and Fukushima and release vast amounts of dangerous radioactivity. There have been 99 nuclear accidents from 1952 to early 2010, totalling $20.5 billion in damages worldwide (Sovacool 2011).

- When nuclear power stations reach the end of their design life, they are so radioactive that dismantling them is a problem. Demolition waste must be stored for thousands of years (Wald 2009). Costs of decommissioning could be as high as the original capital cost of building the nuclear power station in the first place (Diesendorf 2014).

Any one of the above points should suggest we avoid nuclear power. Taken together they emphatically tell us that nuclear power is not going to help heal the world. It will just make things worse. It is part of the high-tech ideology of 'High Energy Earth' (BI 2013). Such an approach derives from a fantasy fuelled by the belief we should have a nuclear-powered *Star Trek* future. In fact it is likely to lead to far greater problems and quite possibly lead to a nuclear war that will wound the world even more. The latest part of this fantasy is that we are told we will move to supposedly 'clean' nuclear *fusion* technology. However, apart from the fact that this has never operated effectively commercially, the high energy neutrons produced by fusion would almost certainly mean that a fusion plant would likely be used as a 'fast breeder' to create fuel for fission plants (Manheimer 2009). Fusion will thus not get rid of dangerous fission plants, but most likely would expand their use. It would then create a plutonium economy, along with the likely proliferation of nuclear weapons (Lidsky 1983; Goldston et al. 2009). This would of course make nuclear terrorism more likely.

Conclusion

It is no exaggeration to say that whether we can heal our world will depend on how we as a society deal with technology. Like all tools, technology can be used for 'good' or 'ill'. We should operate all technology with humility and according to the 'Precautionary Principle' (Washington 2015). One thing governments can do is to establish bodies like the former US Congress Office of Technological Assessment (www.princeton.edu/~ota/). This provided accurate assessments of science and technology, something most politicians are largely uneducated about. If we step away from techno-centrism, from modernism and resourcism, from the 'Mastery of Nature' ideology, then we can intelligently and eco-ethically use the best forms of technology to aid us to heal the world. Energy efficiency, energy conservation, renewable energy, technologies to dematerialise and support eco-design and eco-effectiveness are 'appropriate' (Ibid.). Nuclear power, geoengineering, and carbon capture and storage are not. All technology is not evil, but the rampant use of technology without an 'Earth ethics' (that respects people *and* nature) has led us to the brink of disaster (Ibid.). Just as a fire can be your friend and warm your home and cook your food, it can also burn down your house and community. If we are to heal our world, then we must use appropriate technologies to live respectfully with the rest of life, within the ecological limits of the Earth. That means we should encourage helpful technologies and abandon dangerous and inappropriate ones. Renewable energy is the great win/win solution of our times. It stops the production of greenhouse gases and avoids the need to consider dangerous nuclear technology. However, even with renewable energy we must accept we cannot keep increasing our energy use *forever*. We need to stabilise and then reduce overall energy use to heal our world (Washington 2015).

The most frustrating aspect of the whole technology debate is that many governments and corporations remain in denial about this, and do all they can to oppose and hold back the flowering of renewable energy. However, this flowering of renewables is a major part of healing the world. We must rapidly remove the denial and opposition about renewables and transform our society so that it is powered by 100% renewables within 20 years (Washington 2015). Unconstrained and rampant use of technology helped create the environmental crisis.

However, thoughtful, appropriate and sustainable technology can help get us out of our predicament and heal the world. To heal the planet we need to accept that technology must be guided by an ecocentric worldview and an Earth ethics (see Chapter 2). Then it can be truly appropriate to heal our problems.

What can *I* do?

There are many things one can do to promote *appropriate* technology, being:

1) *Talk about appropriate technology.* Our society is quite drunk on technology, with little questioning of its impacts, or any consideration of its merit as a real solution to our predicament. Point out that all technology is not the same and not uniformly good.
2) Speak out against the glibness of techno-centrism, techno-optimism and ecomodernism, which deny ecological limits and promise a rosy techno-logical future *without* actually addressing the key drivers of unsustainability (overpopulation, overconsumption and the endless growth economy).
3) Champion appropriate technologies such as renewable energy, energy effi-ciency and energy conservation.
4) If you are in a position to put renewable energy on your dwelling, then install this. Aim also for a low energy, carbon-neutral home. Improve your home's energy efficiency (get an energy audit to assist you).
5) Also talk about *in*appropriate technologies that hinder healing our world. In particular, challenge the advertising and PR put forward by nuclear lobbyists that seek to push society into an unsustainable nuclear future.

Note

1 'Scientism' can be described as the belief that the methods of natural science, or the categories and things recognised in natural science, form the only proper elements in any philosophical or other inquiry (Blackburn 2005: 331–2). It thus ignores the rele-vance of worldview and ethics.

References

Alamgir, M., Campbell, M., Sloan, S., et al. (2019) 'High-risk infrastructure projects pose imminent threats to forests in Indonesian Borneo', *Nature*, **9**: 140. DOI: 10.1038/s41598–018-36594-8.

Alexander, S. (2018) 'Home biogas: Turning food waste into renewable energy', *The Con-versation*, January 12, see: https://theconversation.com/home-biogas-turning-food-waste-into-renewable-energy-89920 (accessed 4 May 2019).

Arsenault, C. (2017) 'Only 60 years of farming left if soil degradation continues', *Scientific American*, 4 December 2017, see: www.scientificamerican.com/article/only-60-years-of-farming-left-if-soil-degradation-continues/. (accessed 20 March 2019).

Baxter, T. (2017) 'It's time to accept carbon capture has failed – here's what we should do instead', *The Conversation*, 24 August 2017, see: https://theconversation. com/its-time-to-accept-carbon-capture-has-failed-heres-what-we-should-do-instead -82929. (Accessed 17 March 2019).

BI (2013) 'Embracing our high energy planet'. Breakthrough Institute, see: https://theb reakthrough.org/issues/conservation/embracing-our-high-energy-planet. (accessed27 March 2019).

BI. (2018) 'Ecomodernism 2018: Achieving disagreement (agenda)', see: https://thebreak through.org/events/ecomodernism-2018-achieving-disagreement/agenda   (accessed 15 March 2019).

BI (n.d.) 'Nuclear for 1.5 degrees', *Breakthrough Institute*, see: https://thebreakthrough.org /issues/energy/nuclear-for-1-5-c. (accessed 27 March 2019).

Blackburn, S. (2005) *The Oxford Dictionary of Philosophy*, Oxford: Oxford University Press.

Boyden, S. (2004) *The Biology of Civilisation: Understanding Human Culture as a Force in Nature*, Sydney: UNSW Press.

Brown, L. (2006) *Plan B 2.0: Rescuing a Planet under Stress and a Civilization in Trouble*, New York: W.W. Norton and Company.

Brown, L. (2011) *World on the Edge: How to Prevent Environmental and Economic Collapse*, New York: W.W. Norton and Co.

CIEL (2019) *Fuel to the fire: How geoengineering threatens to entrench fossil fuels and accelerate the climate crisis, Center for International Environmental Law*, see: www.ciel.org/reports/fuel-to-the-fire-how-geoengineering-threatens-to-entrench-fossil-fuels-and-accelerate-the-climate-crisis-feb-2019/. (accessed 17 March 2019).

Curry, P. (2011) *Ecological Ethics: An Introduction*, 2nd ed. Cambridge: Polity Press.

Daly, H. (1991) *Steady State Economics*, Washington: Island Press.

Diesendorf, M. (2007) *Greenhouse Solutions with Sustainable Energy*, Sydney: UNSW Press.

Diesendorf, M. (2013) 'The economics of nuclear energy', in *Nuclear Power and Energy Security in Asia*, eds. R. Basrur and C. Koh, London: Routledge, pp. 50–71.

Diesendorf, M. (2014) *Sustainable Energy Solutions for Climate Change*, Sydney: NewSouth Publishing and Abingdon UK: Routledge-Earthscan.

EM. (2015) 'The Ecomodernist Manifesto', see: www.ecomodernism.org/manifesto-english/.

GDRC. (n.d.) 'Urban and ecological footprints', *Global Development Research Center*, see: www.gdrc.org/uem/footprints/index.html.

GEA. (2012) *Global Energy Assessment - toward a Sustainable Future*, Cambridge, UK: Cambridge University Press, see: www.globalenergyassessment.org (accessed 13 March 2019).

GFN. (2015) 'How can Mediterranean societies thrive in an era of decreasing resources?' *Global Footprint Network*, see: www.footprintnetwork.org/content/documents/ MED_2015_English.pdf. (accessed 27 March 2019).

Goldston, R. J., Glaser, A. and Ross, A. F. (2009) 'Proliferation risks of fusion energy: Clandestine production, covert production, and breakout'. *In 9th IAEA Technical Meeting on Fusion Power Plant Safety*. see: http://web.mit.edu/fusion-fission/HybridsPubli/ Fusion_Proliferation_Risks.pdf. (accessed 15 March 2019).

Guardian. (2018) 'Drilling starts to tap geothermal power from Cornwall's hot rocks', *The Guardian* 7 November 2018, see: www.theguardian.com/business/2018/nov/06/drilling-starts-to-tap-geothermal-power-from-cornwalls-hot-rocks. (accessed 15 March 2019).

Guo, J., Liu, X., Zhang, Y., et al. (2010) 'Significant acidification in major Chinese croplands', *Science*, **327**: 108–110.

Hoggan, J. (2009) *Climate Cover Up: The Crusade to Deny Global Warming*, Vancouver: Greystone Books.

IRENA (2018) 'Renewable energy power costs in 2017⊠, International Renewable Energy Agency, see: www.irena.org/-/media/Files/IRENA/Agency/Publication/2018/Jan/IRENA_2017_Power_Costs_2018.pdf (accessed 15 March 2019).

Jacobson, M. and Delucchi, M. (2011) 'Providing all global energy with wind, water, and solar power, Part I: Technologies, energy resources, quantities and areas of infrastructure, and materials', *Energy Policy*, **39**: 1154–1169.

Jacobson, M., Delucchi, M., Cameron, M., et al. (2018) 'Matching demand with supply at low cost in 139 countries among 20 world regions with 100% intermittent wind, water, and sunlight (WWS) for all purposes', *Renewable Energy*, **123**: 236–248.

Kumar, A. and Ramana, M. (2009) 'The safety inadequacies of India's fast breeder reactor', *Bulletin of the Atomic Scientists*, 21 July 2009.

Laurance, W. F., Burgess, N. D., Hatchwell, M., et al. (2018) 'Towards more sustainable infrastructure: Challenges and opportunities in the ape-range states of tropical Africa and Asia', in *State of the Apes: Infrastructure Development and Ape Conservation*, eds. H. Rainer, A. White and A. Lanjouw, Cambridge, UK: Cambridge University Press and Arcus Foundation, pp. 10–39.

Lidsky, L. M. (1983) 'The trouble with fusion'. *MIT Technology Review*, October 1983: 1–12. See: http://orcutt.net/weblog/wp-content/uploads/2015/08/The-Trouble-With-Fusion_MIT_Tech_Review_1983.pdf. (accessed 15 March 2019).

Louv, R. (2005) *Last Child in the Woods: Saving Our Children from Nature-Deficit Disorder*, London: Atlantic Books.

Lovins, A. (2006) *Quoted in WADE (2006) World Survey of Decentralised Energy 2006*, Washington, DC: World Alliance for Decentralised Energy, see: www.networkforclimateaction.org.uk/toolkit/positive_alternatives/energy/WADE_-_World_Survey_of_Decentralized_Energy06.pdf (accessed 22 March 2019) .

Makhijani, S. and Ochs, A. (2013) 'Renewable energy's natural resource impacts', in *State of the World 2013: Is Sustainability Still Possible?*, ed. L. Starke, Washington: Island Press, pp. 84–98.

Manheimer, W. (2009) 'Hybrid fusion: The only viable development path for tokamaks?', *Journal of Fusion Energy*, **28**: 60–82.

Mason, M. (2012) *Environmental Democracy*, London: Earthscan Publications.

Meadows, D., Randers, J. and Meadows, D. (2004) *Limits to Growth: The 30-year Update*, Vermont: Chelsea Green.

Meyer, W. (2013) *The Environmental Advantages of Cities: Countering Commonsense Antiurbanism*, Cambridge, MA: MIT Press.

Mudd, G. and Diesendorf, M. (2008) 'Sustainability of uranium mining and milling: Toward quantifying resources and eco-efficiency', *Environmental Science & Technology*, **42** (7): 2624–2630.

Nicholson, S. (2013) 'The promises and perils of geoengineering,' in *State of the World 2013: Is Sustainability Still Possible?* ed. L. Starke, Washington: Island Press, pp. 317–331.

Pittock, A. B. (2009) *Climate Change: The Science, Impacts and Solutions*, Australia: CSIRO Publishing/Earthscan.

REN21 (2018) Renewables 2018 Global Status Report, Renewable energy policy network for the 21st century, see: www.ren21.net/wp-content/uploads/2018/06/17-8652_GSR2018_FullReport_web_final_.pdf. (accessed 20 March 2019).

Robock, A. and Toon, B. (2010) 'South Asian threat? Local nuclear war = global suffering', *Scientific American*, **302**(1): 74–81.

Rolston III, H. (2012) *A New Environmental Ethics: The Next Millennium of Life on Earth*, London: Routledge.

Schumacher, E. F. (1973) *Small Is Beautiful: A Study of Economics as if People Mattered*, New York: Blond and Briggs.

Sovacool, B. (2011) *Contesting the Future of Nuclear Power: A Critical Global Assessment of Atomic Energy*, Singapore: World Scientific Publishing Co., see: www.worldscientific.com/worldscibooks/10.1142/7895 (accessed 22 March 2019).

Staples, W. and Cafaro, P. (2012) 'For a species right to exist', in *Life on the Brink: Environmentalists Confront Overpopulation*, eds. P. Cafaro and E. Crist, Georgia, USA: University of Georgia Press, pp. 283–300.

Steffen, W., Richardson, K., Rockstrom, J., et al. (2015) 'Planetary boundaries: Guiding human development on a changing planet', *Science*, **347**(6223). DOI: 10.1126/science.1259855.

Thomasberger, A. (2019) 'Strengthening traditional environmental knowledge for the integration of social and ecological justice', in *Conservation: Integrating Social and Ecological Justice*, eds.. H. Kopnina and H. Washington, New York: Springer, pp. 121–138.

Trembath, A. (2013) 'Embracing our high energy planet', *Breakthrough Institute*, see: https://thebreakthrough.org/issues/conservation/embracing-our-high-energy-planet.

UCS. (n.d.) 'Environmental impacts of hydroelectric power', *Union of Concerned Scientists*, see: www.ucsusa.org/clean_energy/our-energy-choices/renewable-energy/environmental-impacts-hydroelectric-power.html. (accessed 12 March 2019).

Wald, M. (2009) 'Dismantling nuclear reactors', *Scientific American*, January 26, 2009.

Washington, H. (2015) *Demystifying Sustainability: Towards Real Solutions*, London: Routledge.

Washington, H. (2018) *A Sense of Wonder Towards Nature: Healing the World through Belonging*, London: Routledge.

Washington, H. (2019) unpublished poem 'Appropriate'.

Wilson, E. O. (2016) *Half-Earth: Our Planet's Fight for Life*, New York: Liveright/Norton.

WWF. (2011) *The Energy Report: 100% Renewable Energy by 2050*, Switzerland: WWF.

WWF. (n.d.) 'Tree clearing', *World Wide Fund for Nature*. see: www.wwf.org.au/what-we-do/species/tree-clearing#gs.Ac80uJGk (accessed 2 March 2019).

WWI. (2016) *Can a City Be Sustainable: State of the World Report*, *Worldwatch Institute*, Washington: The Worldwatch Institute.

Yigitcanlar, T., Foth, M. and Kamruzzaman, M. (2018) 'Towards post-anthropocentric cities: Reconceptualising smart cities to evade urban ecocide', *Journal of Urban Technology*. DOI: 10.1080/10630732.2018.1524249.

7

REINVENT AND REJUVENATE YOURSELF

Reject hubris

Could I but touch
The source of soul
The spirit centre
The perfect place,
That place of peace.
There'd be no fear
Fear of unloved,
There'd be no trace
Of human bitterness.
Reach down inside,
Touch with your hands
The harmony point
Where you are one:
Entire, whole.

'The Perfect Place' (Washington 2010)

SUMMARY

The solutions to our predicament have been known for decades, yet society has not acted. Part of the problem is that many people in society have been psychologically damaged, making it harder for them to become active to heal the world. Hence we need to reinvent or heal ourselves. We need to rethink who we 'are' as opposed to who society or corporations want us to be. We need to re-examine our ethics, our own denial and our own connection to nature (our eco-spirituality and sense of wonder). We also need to reject the 'hubris' or excessive pride present in our society, which is based on the idea that we can make and 'control' the world (through being 'masters'). For the detail explaining this summary, please read on; however, if you just want to know the things *you can do* (six are listed) about this, then turn to the end of the chapter.

Introduction

The solutions to the environmental crisis have been known at least since William Catton wrote his landmark book *Overshoot* in 1982. So why hasn't society acted on these problems and solved them? Well, other chapters come up with various reasons. One obvious reason is that we deny we have any problems, or we pretend that out technology will solve them all. However, there is also the aspect of apathy, where we may know that things are wrong, but we never get around to doing anything about acting to solve them. Part of the problem is that most of us carry psychological scars; we have been pushed into a mould to conform to what parents and society *want* us to be. Peer pressure is a powerful thing (Washington 2015). Thus we are not in fact 'ourselves', or what we would have been had we been nurtured to develop into our real selves. Accordingly, many of us are psychologically and spiritually damaged, and that damage makes it harder for us to act. Yet we need to act if healing is to take place. That means we must also *heal ourselves* to heal the planet (Sattman-Frese and Hill 2008). I am indebted to the work of social ecologist Prof Stuart Hill of Western Sydney University for bringing this fully to my attention.

What damages us psychosocially?

It has been pointed out that our species is psychosocially undeveloped, that the path to sustainability will also require personal transformation and development, and a cultural or psychosocial evolution within most societies (Hill 1991, 2014). This is possible, but we need to recognise the need for such evolution as a key part of the solution (Ibid.). Most of us have grown up within a society that is unsustainable, one based on an idea of endless growth, competition and inequality. Accordingly, most of us have (to a greater or lesser extent) been conditioned or damaged within our inner selves. Children may often be raised (mostly unintentionally) in oppressive ways by parents and others (who were themselves damaged by being raised that way). Hence most of us suffer a deep inner distress. However, for the first time in history we have the means to develop psychosocially, and Hill argues we have the theory and practice to recover from distress, and that this is necessary if we are to reach sustainability (Hill 1991). This is why we need to redesign ourselves so we can heal the planet.

We will have to pay more attention to how we raise and educate our children, and rethink dubious social attitudes towards 'worth' and 'status', and rethink the design of institutional structures and processes. Our institutions should support the healing of our world rather than remain a barrier (Hill 2014). Hill (2014) argues that we need to build our 'personal capital' or personal sustainability to enable us to have the inner strength to create real change. This means empowerment, awareness, creative visioning, worldview clarification and developing our respon- sibility and spontaneity (Ibid.). We need to develop our ability to care, feel empathy, love, be mutualistic, value equity and to extend our compassion to other species and the planet (Hill 2014; see also Chapter 2). We need to support our personal development and continue a process of life-long learning. This will enable us to build our social capital. We will also need to develop our inter- cultural or 'inter-personal capital' to build trust and cooperation. We will need to be 'proactive' rather than reactive, and accept the need to fundamentally reinvent our society, rather than just 'increase efficiency' (Hill 2014). Too much of the dis- cussion and action around 'sustainability' has really just continued the 'business-as- usual' approach (Washington 2015). We also need to put 'change' in perspective. Most of the changes towards sustainability we can make individually will realistic- ally be small, but they can still be very meaningful. We should celebrate these changes publicly, so they can be copied by others (Hill 2001). Big projects (e.g. giant dams) may be 'newsworthy', but are not necessarily the best action towards sustainability. Real sustainability, and the ability to heal our world, will come from the net result of a great many of us *acting collectively* for change.

BOX 7.1 'WISDOMS' TO ENABLE CHANGE, STUART B. HILL

There are many 'wisdoms' we can adopt to change both ourselves and the world:

- Act on 'love' vs 'fear'.
- Plan and implement 'small meaningful initiatives that you can guaran- tee to carry through to completion' – and enable them to become 'contagious' (through publicity and communication) – and progress from such actions to the next meaningful actions.
- Engage with others in understanding, planning and action – we are a social species! – individual action is often subconsciously contamin- ated by fears and long-held 'compensatory behaviours', such as want- ing to do things alone, to appear powerful, to compensate for adaptive historical disempowerment.
- Ask of all theory and practice – what is it in the service of? – before supporting or copying it.
- Collaborate across difference to achieve broadly shared goals – don't end up isolated, alone in a 'sandbox'.
- To achieve sustainable progressive change, focus (at least first) on enab- ling and collaborating with the 'benign' agendas of others vs. trying to impose on them your own 'benign' agendas.

- Focus on enabling the potential of people, society and nature to express itself – so that wellbeing, social justice and sustainability can emerge (in integrated, synergistic ways).
- Outcomes are only as good and sustainable as the people creating and implementing them – so start with the people.
- Celebrate publicly at every opportunity – to enable the good stuff to spread and become 'contagious'.
- Use the media – let me repeat – use the media! – such 'political' communication is key to enabling change.
- Keep working on and implementing – especially with others – your (shared) benign visions (this takes 'stickability').
- Work sensitively with time and space, especially from the position of the 'others' (ask: who, what, which, where, when, how, why, if and if not?).
- See no 'enemies' – recognise such 'triggers' as indicators of woundedness, maldesign and mismanagement – everyone is always doing the best they can, given their potential, past experience and the present context – these are the three areas to work with when enabling 'progressive' change.
- Don't get stuck in endless 'measuring studies' ('monitoring our extinction') – these are often designed to postpone change that is perceived as threatening to existing power privileges and structures.

For more information: See: www.stuartbhill.com/.

Rethinking our ethics

Reinventing ourselves will require many of us to radically rethink our worldview and ethics (see Chapter 2). I remember teaching a class on Green Politics at Western Sydney University, and asking my class: 'Does nature have intrinsic value, a right to exist for itself?' I got some blank looks, but later a student came up to me with a very worried look and asked: 'Why have I never thought about this? Why have I never been asked to think about it?' I thought for a moment and said: 'They are good questions, aren't they?' They were indeed good questions, because they showed how deeply anthropocentric our society remains. In Western society, the idea that nature has intrinsic value is often not even raised as a possibility for discussion. Many scientists still maintain that science is not about ethics; however, today it has to be (Washington 2018a). For a start, the way scientists are taught is not ethics free; they are bombarded by anthropocentric and neoliberal ideology (Weber 2016; Washington 2018a). Many of the commonly accepted terms of ecology (for example) actually have anthropocentric roots, such as 'adaptive management', 'resilience thinking' and 'ecosystem services' (Washington 2018b, 2019). Even worse, the extent of anthropocentrism and neoliberalism seems to be becoming worse, with the promotion of anthropocentric ideologies such as 'new conservation', 'critical social science', ecomodernism and ecosystem services (Monbiot 2014; Kopnina et al. 2018; Kopnina and Washington 2019). If we are going to reinvent ourselves we will essentially have to reconsider whether we have been brainwashed into thinking overwhelmingly along the lines of anthropocentric and neoliberal ethics.

Rethinking who we are

It may sound strange, but we will need to rethink who 'we' *are*. If we have been brainwashed into thinking anthropocentrically (and along neoliberal lines), then perhaps we have been brainwashed into other things? Perhaps we have been brainwashed into thinking that 'growth is always good'? Brainwashed into thinking technology can do anything? Brainwashed into 'Shop till you drop!' so that rampant consumerism seems normal? Brainwashed into thinking that all of existence is just a 'resource' for our use? Brainwashed into thinking that we philosophically 'own' all of existence, which we must then dominate and control? Brainwashed into denial and apathy? Alan Watts (1973) argued there is a mighty taboo in society – our tacit conspiracy to ignore who, or what, we really *are*. He argues (Watts 1973: 9):

> [t]he prevalent sensation of oneself as a separate ego enclosed in a bag of skin is a hallucination which accords neither with Western science nor with the experimental philosophy-religions of the East – in particular the central and germinal Vedanta philosophy of Hinduism. This hallucination underlies the misuse of technology for the violent subjugation of man's natural environment and, consequently, its eventual destruction.

He argues (p. 8) that we urgently need: 'a sense of our own existence which is in accord with the physical facts and which overcomes our feeling of alienation from the universe'. Watts (1973: 87) goes on to describe the problems of many who are locked into Western thought:

> But the question is how to get over the *sensation* of being locked out from everything 'other', of being only oneself – an organism flung into unavoidable competition and conflict with almost every 'object' in its experience.

Hence, part of rejuvenating yourself is seeking to know who you truly are. Not who society or corporations want you to be (a consumer), but (hopefully) an eco-citizen who sees herself as part of the rest of life. We need to question the teachings that have indoctrinated us to think we were locked out from everything 'other', and realise that we are kin with the rest of life. That means adopting an ecocentric worldview and an ecological ethics (see Chapter 2).

Rethinking our own denial

Chapter 3 looks at the need to accept reality and break the denial dam. Rejuvenating ourselves will definitely mean examining and accepting one's *own* denial. This is harder than it sounds. We often say 'we learn from our mistakes', but actually many of us refuse to admit that we ever made any. Tavris and Aronson (2007) point out the problem of 'cognitive dissonance', which is the rather sick feeling you get when you realise a cherished belief is not supported by the evidence. They explain that most people will go to great lengths to reduce cognitive dissonance. Quite commonly, that means denying the evidence that caused the dissonance (Ibid.). Hence people can deny scientific evidence that shows there is an environmental crisis. British politician Lord Molson famously stated: 'I will look at any additional evidence to confirm the opinion to which I have already come' (in Tavris and Aronson 2007: 17). It thus takes a certain degree of 'big-heartedness' to admit to one's own denial. Accordingly, a key part of redesigning oneself so one can heal has to be analysing one's own denial – and taking steps to move past it.

Rethinking our eco-spirituality and sense of wonder

Rejuvenating ourselves will mean thinking about our 'spiritual' capital as much as our built and human capital. Surprisingly, many people react negatively to the term 'spirituality' (seemingly because they mistakenly equate it with religion). However, 'spirituality' is really no more than suggesting (among other things) that the whole is greater than the sum of its parts. As Watts (1973) noted, knowing who you 'are' is in large part breaking down the idea that we are 'Masters of the Universe' totally different from the rest of nature. We are only separate if we persist in thinking that we are; and rethinking eco-spirituality is about recognizing that we are part of nature, that she is our kin (Washington 2018a). This requires rejuvenating our sense of wonder, which is discussed in Chapter 9.

Nature rituals such as the 'Council of All Beings'

An important part of rethinking our eco-spirituality is taking part in 'nature rituals' such as deep ecology workshops, in particular the 'Council of All Beings' (Washington 2018a). John Seed (RIC 2001) is a deep ecologist who has carried out such workshops for some decades, and states:

> If we look at indigenous cultures, we may notice that without exception rituals affirming and nurturing the sense of interconnectedness between people and nature play a central role in the lives of these societies. This suggests that the tendency for a split to develop between humans and the rest of nature must be very strong. Why else would the need for such

rituals be so universally perceived? It also suggests the direction we must search for the healing of the split: we need to reclaim the ritual and ceremony which were lost from our culture a long time ago, and to our amazement we find that this is incredibly easy to do.

Ritual and ceremony have always been important for connecting and reconnecting people to the land. It is also important in how society makes decisions regarding the land (Zimmerman and Coyle 1996). Ritual forms the basis of many forms of connection to nature – practical, ethical and spiritual. Ritual is also a key part of most religions and spiritual practices (Taylor 2010). It is clearly something through which humanity seeks to express what, at its simplest, is a giving of thanks to life – for living in such a beautiful and mysterious world (Washington 2018a). One important nature ritual is the 'Council of All Beings' used in deep ecology workshops (RIC, 2001; Macy and Brown 2014). The process is summarised by Washington (2018a):

1) The guide generally asks people to go outside into nature for a certain time (perhaps 10–20 minutes) to discover their 'ally' in nature. As Macy (in Taylor 2008: 427) notes:

    Participants begin by letting themselves be chosen by another life-form, be it animal, plant, or natural feature like swamp or desert. … go with what first intuitively occurs to them, rather than selecting an object of previous study. … Then we take time to behold this life-form in our mind's eye, bestowing upon it fullness of attention, imagining its rhythms and pleasures and needs. Respectfully, silently, we ask its permission to speak for it in the Council of All Beings.

2) Participants then return and make simple masks, working together in companionable silence with paper and paints, twigs and leaves. Then, briefly clustering in small groups, they practice *taking on the identity* of their chosen being. This helps participants let go of their self-consciousness as humans, and become more at ease in imagining a very different perspective on life. Then, with due formality, the participants assemble in a circle and the Council commences. Macy (in Taylor 2008: 427) notes that:

    When I am the guide, and speaking, of course, as my adopted life-form, I like to begin the proceedings by inviting the beings to identify themselves in turn, a kind of roll call: 'Wolf is here, I speak for all wolves'; 'I am Wild Goose; I speak for all migratory birds'.

Seed (RIC n.d.) notes the guide can explain: 'We come to this Council to share matters that are close to our hearts, and also to share our strengths, our beauty, our troubles and our wisdom'.

3) Welcoming them all, the guide thanks them for coming and, with some solemnity, sets the theme for Council deliberations. This may be: 'We meet in Council because our planet is in trouble; our lives and our ancient ways are endangered. It is fitting that we confer, for there is much now that needs to be said and much that needs to be heard' (Macy n.d.). However, Macy and Seed both make it clear it could really be on *any* topic that concerns the living world.

4) The Council unfolds in three consecutive stages. First, the Beings address each other, telling of the changes and hardships they are now experiencing. The second stage of the Council begins after most have spoken, and the guide *invites humans* into the centre. Macy and Brown (2014) note that since it is clear that one young species is at the root of all this trouble, its representatives should be present to hear these testimonies. So, a few at a time, the Beings put aside their masks and move to sit for a while, as humans, in the middle of the circle. The other life-forms/beings in the circle now speak to them directly (the humans remain silent). For example, Macy (n.d.) notes:

> For millions of years we've raised our young, rich in our ways and wisdom. Now our days are numbered because of what you are doing. Be still for once, and listen to us.

5) Most of the humans in the centre now return to the circle and put on their masks. In the third stage of the Council, the other life-forms *offer gifts* to the humans. Recognizing their kinship with humankind, they would help this young species deal with the crisis it has created. With this invitation, the beings in the Council begin spontaneously to offer their own particular qualities and capacities. Macy and Brown (1998: 164) suggest examples such as:

> As Mountain I offer you my deep peace. Come to me at any time to rest, to dream. Without dreams, you may lose your vision and your hope. Come, too, for my strength and steadfastness whenever you need.

Humans are allowed only to say 'thank you'.

6) One after another the Beings offer their particular powers to help the humans in the centre. After speaking, each leaves its mask in the outer circle and joins the humans in the middle, receiving the gifts still to be given.

7) Macy (n.d.) notes: 'These gifts reside already in the human spirit, as seeds within the psyche; otherwise they could not be spoken. Their naming brings forth a sense of wholeness and glad possibility'. When all of them have been offered, the Council of All Beings is formally concluded.

8)   Macy (n.d.) suggests that care is taken to *thank the life-forms* (or other beings), who have spoken through participants and to dispose of the masks in a deliberate fashion. The masks may be formally burned, or hung on a tree or wall, or taken home with participants as a symbolic reminder of the ritual.

9)   The Council may end with a reading (e.g. from 'Thinking like a Mountain', Seed et al. 1988) or a chant.

Macy and Brown (2014: 164) advise participants to *thank their Being* and then: 'let the identity go'. They should kneel and bow with hands and head touching the ground. This ritual may be shortened by not moving formally to the second and third stages, as aspects of these may appear anyway (this is Seed's current practice). There are many other nature rituals used in deep ecology workshops. You can find out more about various nature rituals that are part of a deep ecology workshop at www.rainforestinfo.org.au/deep-eco/welcome.htm, and in Macy and Brown (2014). There are also other nature rituals such as the 'Rite of Passage' for adolescents, 'yatras' (journeys to sacred places) and vision quests (Washington 2018a). Other rituals could include green weddings and green funerals, a ritual for assuming custodianship (not ownership) for an area of land, and rituals for seasonal change.

Rejecting hubris

A part of reinventing ourselves is the rejection of *hubris*, of the arrogant pride that assumes we are 'Masters of the Universe'. We should not operate from hubris when dealing with 'persons', whether they are human persons or nonhuman persons (as we should think of the rest of life as being, Harvey 2005). Instead, we need to live and work from a position of humility. Neither humanity as a whole, nor you or I are the 'be all and end all' of things. Compared to the world or the Universe, we are just a tiny mote, not insignificant (for we are part of life) but neither of overwhelming consequence. We need to regain a sense of perspective. This is *not* the human planet, nor solar system, nor galaxy. Rather, the Earth is the ground of our being, and equally importantly, the ground of *all* other living beings. Our dealings with people and nature should thus be humble, not based on arrogance, self-importance or narcissism. Neither we (nor all humanity) 'can do anything we want', though this is a common aspiration of so many self-help books and videos. Not being able to 'do anything' is both a practical reality and an ethical principle. Hubris is part of the insane idea that we can dominate *everything*, both human and nonhuman. Hubris is a key aspect of any megalomania. We should understand that both physically and ethically we cannot 'do anything', nor should we try to. We should do what is *right*, enable healing of life, and not undermine and destroy it. As a society, our attempts to do whatever we want have created economies and societies that are fundamentally unsustainable. These are in the process of bankrupting nature (Wijkman and Rockstrom 2012). Rejecting hubris is part of becoming sane and respectful of the rest of life, it is relearning a kinship ethics so that we can love the rest of life that is our kin. With hubris we cannot heal; with humility towards all life – we can.

Conclusion

Rejuvenating yourself may sound strange at first glance, but it recognises that healing the world may first involve healing yourself. Watts (1973) suggests that a first step to this is getting to know who you truly 'are'. Other steps involve thinking deeply about your own ethics, your patterns of denial and about your eco-spirituality. An important aspect of this is taking time to be present in wild nature, which is a very healing place. Indeed, it has been said that wilderness and wild places are worth thousands of beds in a mental hospital (Nash 1967). From decades of my own experience, wilderness and wild places are also clearly a great place to gain perspective on our society, and think about what needs to be supported and healed (Washington 2018a). It is that deep connection to nature, that sense of wonder, that can help heal us so we can heal the world. This is discussed further in Chapter 9.

What can *I* do?

Here are several things that each of us might choose to do to reinvent and rejuvenate ourselves:

1) Accept that you might need some personal healing to enable you to be better able to heal the world. Many of us (myself included in the past) tend to operate from the mindset of 'I can't worry about such touchy-feely stuff; I need to get on with practical things!' I did this for 20 years, and then burnt out for a while – *then* I had to heal myself.
2) Stop and think about what you might need to rejuvenate in yourself. Have you been damaged by things you would rather not think and talk about? How well do you really know who you actually are? What are the things you might be in denial about? Most of us have a range of them. What would it take for you to think (and feel) more deeply about your own philosophy and ethics (see Chapter 2)? Also think about your own eco-spirituality and sense of wonder (see Chapter 9). In what ways might you be stuck in hubris and arrogance about humanity's ability to fix everything technologically (see Chapter 6)? How might you benefit from working on issues of self-empowerment, awareness, creative visioning, worldview clarification and developing one's responsibility and spontaneity (Hill 2014)?
3) One thing you can do is take part in nature rituals such as deep ecology workshops, to reconnect with nature. One of the most effective rituals is the 'Council of All Beings', where one represents another part of nature

(animate or inanimate). Don't avoid this because you think it feels a bit 'hippy'; dare to *try it out*. Everyone gets something from such rituals. Other rituals you could try are ˌyatras and vision quests (Washington 2018a).

4) Seek to develop your ability to care, feel empathy, love, be mutualistic, value equity and to extend one's compassion to other species and the planet (Hill 2014).

5) Peer pressure is a powerful thing in our society. Examine the ways you might have been brainwashed and nudged into accepting ideologies that needed to be *questioned* and *resisted* (such as human supremacy, growthism, modernism).

6) Look for any ways you might have fallen prey to using avoidance responses, such as 'What is the use?', 'It's too late!', and other personally protective and defensive strategies. It is only too late if we, in fact, do nothing. As Lester Brown (2006) has rightly said 'Saving our civilization is not a spectator sport'.

References

Brown, L. (2006) *Plan B 2.0: Rescuing a Planet under Stress and a Civilization in Trouble*, New York: W.W. Norton and Company.

Catton, W. (1982) *Overshoot: The Ecological Basis of Revolutionary Change*, Chicago: University of Illinois Press.

Harvey, G. (2005) *Animism: Respecting the Living World*, London: Hurst and Co.

Hill, S. B. (1991) 'Ecological and psychological prerequisites for the establishment of sustainable agricultural communities', In *Agroecosystems and Ecological Settlements*, eds. L. Salomonsson, E. Nilsson and T. Jones, Uppsala: Swedish University of Agricultural Sciences, pp. 28 51.

Hill, S. B. (2001) 'Working with processes of change, particularly psychological processes, when implementing sustainable agriculture', In *The Best of … Exploring Sustainable Alternatives: An Introduction to Sustainable Agriculture*, ed. H. Haidn, Saskatoon: Canadian Centre for Sustainable Agriculture, pp. 125–134.

Hill, S. B. (2014) 'Considerations for enabling the ecological redesign of organic and conventional agriculture: A social ecology and psychosocial perspective', In *Organic Farming, Prototype for Sustainable Agricultures*, eds. S. Bellon and S. Penvern, Dordrecht: Springer, pp. 401–422.

Kopnina, H., Taylor, B., Washington, H. and Piccolo, J. (2018) 'Anthropocentrism: More than just a misunderstood problem', *Journal of Agricultural and Environmental Ethics*, **31** (1): 109–127.

Kopnina, H. and Washington, H., eds. (2019) *Conservation: Integrating Social and Ecological Justice*, New York: Springer.

Macy, J. (n.d.) 'The Council of All Beings', see: www.rainforestinfo.org.au/deep-eco /Joanna%20Macy.htm (note this essay was later published in Taylor, B. (2008) *The Encyclopedia of Religion and Nature: Volume 1*, New York: Bloomsbury).

Macy, J. and Brown, M. (1998) *Coming Back to Life: Practices to Reconnect Our Lives, Our World, Gabriola Island*, Gabriola Island, BC, Canada: New Society Publishers.

Macy, J. and Brown, M. (2014) *Coming Back to Life: The Updated Guide to the Work that Reconnects, Gabriola Island*, Gabriola Island, BC, Canada: New Society Publishers.

Monbiot, G. (2014) 'Put a price on nature? We must stop this neoliberal road to ruin', *The Guardian* 24 July 2014, see: www.theguardian.com/environment/georgemonbiot/2014/jul/24/price-nature-neoliberal-capital-road-ruin (accessed 8 July 2018).

Nash, R. (1967) *Wilderness and the American Mind*, 1st ed. New Haven/London: Yale University Press.

RIC. (2001) 'The Council of All Beings', *Rainforest Information Centre*, see: www.rainforestinfo.org.au/deep-eco/council.htm (accessed17 March 2019).

RIC. (n.d.) 'The Council of All Beings workshop manual', *Rainforest Information Centre* website, deep ecology section, see: www.rainforestinfo.org.au/deep-eco/cabcont.htm.

Sattman-Frese, W. and Hill, S. (2008) *Learning for Sustainable Living: Psychology of Ecological Transformation*, *www.lulu.com*, see: www.slideshare.net/WernerSF/lf-sl-booka4aug08. (accessed 27 March 2019).

Seed, J., Macy, J., Fleming, P. and Naess, A. (1988) *Thinking like a Mountain: Towards a Council of All Beings*, Gabriola Island, Canada: New Society Publishers. see: www.rainforestinfo.org.au/deep-eco/TLAM%20text.htm.

Tavris, C. and Aronson, E. (2007) *Mistakes Were Made (but Not by Me): Why We Justify Foolish Beliefs, Bad Decisions, and Hurtful Acts*, Orlando, Florida: Harcourt Books.

Taylor, B. (2008) *The Encyclopedia of Religion and Nature: Volume 1*, New York: Bloomsbury.

Taylor, B. (2010) *Dark Green Religion: Nature Spirituality and the Planetary Future*, Berkeley: University of California Press.

Washington, H. (2010) *Gift of the Wild: The Nature Poetry of Haydn Washington, lulu.com*, see: www.lulu.com/shop/haydn-washington/gift-of-the-wild/paperback/product-14243912.html. (accessed 27 March 2019).

Washington, H. (2015) *Demystifying Sustainability: Towards Real Solutions*, London: Routledge.

Washington, H. (2018a) *A Sense of Wonder Towards Nature: Healing the World through Belonging*, London: Routledge.

Washington, H. (2018b) 'Harmony – Not theory', *The Ecological Citizen*, **1** (2): 203–210.

Washington, H. (2019) 'Ecosystem Services – A key step forward or anthropocentrism's 'Trojan Horse' in conservation?', in *Conservation: Integrating Social and Ecological Justice*, eds. H. Kopnina and H. Washington, New York: Springer, pp. 73–90.

Watts, A. (1973) *The Book on the Taboo against Knowing Who You Are*, London: Abacus.

Weber, A. (2016) *The Biology of Wonder: Aliveness, Feeling, and the Metamorphosis of Science*, Gabriola Island, BC, Canada: New Society Publishers.

Wijkman, A. and Rockstrom, J. (2012) *Bankrupting Nature: Denying Our Planetary Boundaries*, London: Routledge.

Zimmerman, J. and Coyle, V. (1996) *The Way of Council*, Chicago: Bramble Books.

8

BE IN PLACE, DON'T *OWN* IT

Walking the bush
Brother Bird
And Sister Tree
Reach out to me,
They too **belong***,*
So I listen
To their song:
A living poetry.
Can I ignore
Their right to being?
How can I fail
Just to be just?

From 'Justice' (Washington 2013a)

SUMMARY

This chapter focuses on something many of us tend to overlook, the Western idea that we philosophically 'own' the land. Such an idea is essentially the *slavery of place*, where no matter what has evolved or lives on place, it becomes 'ours' through ownership (and we can sell this property to others). This was/is not the idea of many Indigenous societies, but is an anthropocentric idea from Western society that has now been globalised around the world. Under this ideological blueprint, the world is gridded into portions 'owned' by people, who can do whatever they like to 'their' land. This is a commodified sleight-of-hand, whereby in one stroke the intrinsic value, agency and rights of nature of any area of land are turned just into monetary value to be traded. Indeed, it is now commonly applied to all areas, even the ocean bottom for mining. The chapter points out that legal owner-ship is one thing (we all have to live somewhere), but *philosophical* owner-ship is something very different and is a barrier to healing. Related ideas discussed are those of 'control' and 'mastery' of nature, which are common in Western society. The chapter concludes by looking at the interaction of 'rights of the land' with land rights and the idea we have an obligation *to* the land. It suggests that our obligation to the land and the rights of the land need to be given priority. For the detail explaining this summary, please read on; however, if you just want to know the things *you can do* (five are listed) about this, then turn to the end of the chapter.

Introduction

Perhaps nothing shows the extent of how we have been brainwashed by anthropocentrism as the idea that we think we 'own' the land *philosophically*. To own something in this way is to have full control over it, to be a 'master', while effectively the land becomes a slave. If alternatively we lived respectfully in place with the rest of life, we could not think of ourselves as the philosophical 'owner' or master of the land we live on. Rather, we would be a custodian or steward, or perhaps just a 'carer'. Yet the whole basis of the dominant anthropo-centric worldview of Western society rests on 'ownership' and mastery.

Ownership: the slavery of place

We talk all the time of the land we or others 'own', yet we almost never stop to consider the worldview and ethics behind this. Why should the land *belong* to you or me or any human? Countless other forms of life may live there and have evolved there over many millions of years. Similarly, the geodiversity of the place will have been sculpted over aeons of time (Washington 2018a), yet we dismiss all of this by a simple sleight-of-hand by saying – 'I own it'. Instead of this, I am suggesting here an idea that some may find unthinkable (or at least not previously thought about) – that the land *owns itself*, just as the waters, air, clouds and stars own themselves. The galaxy too owns itself. The idea of philosophical ownership is actually the idea of slavery applied to the nonhuman world. Through 'owning it' the land becomes our slave, without voice, without rights, without agency; it becomes just a dead thing that is 'ours' (Vetlesen 2015). Yet so condi-tioned have we become in Western society to 'land as slave' that most of us never stop to question this odd idea. This is very much an ingrained assumption (see Chapter 5) – one that we do not realise we have been brainwashed to believe. Yet all societies have not fallen into this assump-tion. Tecumseh (the great Shawnee leader in the United States) in 1810 expressed his disgust at the idea of ownership (when urged to sell his land) in these words (Josephy 1995: 311):

> Sell a country! Why not sell the air, the clouds and the great sea, as well as the earth? Did not the Great Spirit make them all for the use of his children?

I should point out that the idea of 'children' and persons in many US Indigenous cultures was not limited to humanity. Harvey (2005) describes animists as people who recognise that the world is full of persons, only *some* of whom are human. Thus Tecumseh was pointing to the odd idea that the Europeans thought land could be 'owned', so why not the sea and the air and clouds? Why not indeed as far as Western society is concerned. Why not the stars also, maybe we can own them as well? Or the whole galaxy, and the other galaxies, even the whole Universe? Once one starts philosophically 'owning things' one can extend the slavery of the nonhuman right up to all existence – unless one is sane. Why do I suggest such an idea of 'owning everything' is not sane? I would respond by asking if it was a healthy idea to think one could own a fellow human being? For centuries society thought it was, then only a couple of centuries ago the ethical ramifications of slavery really hit home – and human slavery was outlawed. Similarly, it is not sane to apply 'ownership' to the rest of existence. It is true that humanity still legally 'owns' countless millions of domestic animals, but laws exist to ensure we cannot do anything we like to them, such as cause them suffering (though admittedly such laws are often broken). I also understand that I *legally* own my dog Tinki, but really she is part of my family. While she cannot speak in words, she nevertheless lets me know what she wants with her eyes and bearing (no mastery there, she is very much my dog-daughter).

Imagine if we freed ourselves of the idea of ownership of our domesticated animals and pets? Imagine if we went even further and freed the lands and waters and seas from being slaves? For some this may be a step too far, yet many Indigenous societies did not feel it was. Black Elk of the Sioux has evoked the unity of life by saying (Neihardt 1972: 1):

> It is the story of all life that is holy and is good to tell, and of us two-leggeds sharing in it with the four-leggeds and the wings of the air and all green things; for these are children of one mother and their father is one spirit.

Luther Standing Bear (1928) of the Dakota told of when a young Dakota man went off on a Vision Quest. During such a quest, one spends several days naked and fasting in the mountains (Washington 2018b). In a prayer to the Universe, the Dakota asked (Knudtson and Suzuki 1992: 212):

> O Wakan-Tanka, grant that this young man may have relatives; that he may be one with the four winds, the four Powers of the world, and with the light of the dawn. May he understand his relationship with all the winged peoples of the air ... Our Grandmother and Mother (Earth) ... this young man wishes to become one with all things ... For the good of all your peoples, help him!

Other animals and aspects of nature were thus not 'things', they were *people*. Standing Bear (1978) explains that the Dakota made sure their children knew that wherever they went, they would be greeted by warm, reassuring presences of local life-forms, geological features and natural forces. These were often as trusted, familiar and communicative with them as members of their families back home. Through this process, wrote Standing Bear (1978: 45), Dakota children came to an early understanding 'that we are of the soil and soil of us', that 'we love the birds and beasts that grew with us on this soil'. He said that a bond existed between all living things because they all: 'drank the same water and breathed the same air' (Knudtson and Suzuki 1992: xxxvi). Accordingly, many Indigenous cultures did not believe that they philosophically *owned* the land or other living things. In fact Aboriginal peoples in Australia had no term for 'ownership' (Graham and Maloney 2019). It is true Indigenous peoples hunted and killed animals and harvested plants for food; however, they then thanked them for those gifts to the hunters and gatherers. This was *respect* and *reciprocity* – never an ownership where the rest of life was just a slave. A deep relationship to the land of course has also been part of sustainable agriculture. Part of this has been the understanding of perspective as to our own 'significance'. Berry (1976: 95) observed re those who gain a holistic perspective:

> Seeing himself as a tiny member of a world he cannot comprehend or master or in any final sense possess, he cannot possibly think of himself as a god.

There is also a key problem here concerning the difference between philosophical 'ownership' (the land being a slave) and 'legal' ownership (which we all have to deal with as we all must live somewhere, and legal ownership is the law of our society). I have legal ownership of my block of land, but philosophically there is no way I believe I actually philosophically 'own' it. I am merely the current custodian and carer who seeks to listen to the land and learn wisdom from it (Washington 2018b). My land clearly *owns itself.*

The 'blueprint' of gridding the world

Eileen Crist (2019) writes passionately against the 'blueprint' of planetary ownership. She argues that this blueprint exists by virtue of broad human consent (at least in Western society, now globalised around the world). Crist does acknowledge that neither the blueprint of planetary ownership, nor the gridding it sponsors are universal to human cultures or an inherent drive of 'human nature'. She calls them an 'Earth regime kit' that Western civilisation has developed, and which today has become globally ascendant. Crist notes that this applies to the seas also:

> Shores have been massively appropriated worldwide – an enabling of the shared blueprint of Earth ownership – and turned into 'beaches' and 'beachfront' properties. Entire regions of the ocean have been configured into 'fisheries', a weird albeit commonplace term that fuses the fish, their places, and the industrial fishing industry into one (presumed good) conglomerate.

She elaborates that even as she writes: 'the seabed is being sliced up by nation-states and allied industry marauders for mining polymetallic nodules, seamounts, and hydrothermal vents'. She notes that the compulsion to affirm human control of geographical space and ownership of Earth's estate is feeding the expansion of the grid. Crist notes that people are socially conditioned throughout their lives to believe this idea, so it gets little publicity. She further notes that: 'Humans control (ultimately all) geographical space and consider themselves entitled to do so, full stop. (This presumption is so normalized that the Moon, Mars, and their "resources" also appear, *naturally*, up for grabs.)'. Crist argues that:

> We are thus called to undo the blueprint both in our minds and in the world. We must unmask its guise as normal and understand Earth's possession for what it is – the exercise of power, violence, and injustice on a cosmic scale.

Crist (2019) speaks about conservation and 'freedom':

> It is about setting Earth free to be an expansive, untamed, and exuberant mandala of life that can *actually*, if implemented in timely fashion, heal many ecological wounds … authentic human freedom can never be founded on annihilating, constricting, and enslaving nonhumans nor can it blossom in the bleak landscape of Earth bondage and ruins. … The prerequisite for realizing authentic human freedom is to free humans from the debasing shackles of human supremacy, lifting humanity into the infinite sight of the fundamental goodness of all life freed.

We should all thus be champions in a movement to free nonhumans from the shackles of ownership and bondage to human supremacy.

'Control' and 'mastery' of nature

Why should Western society seek control and mastery over nature? Some would glibly say: 'It is human nature', but actually this is just a reflection of the current dogma in Western society (Manner and Gowdy 2010). Modernists believe that 'human nature' is purely selfish and greedy. They argue society has always operated on Tennyson's idea (1849) of a: 'nature red in tooth and claw'. However, the concept of 'human nature' is actually dependent on your worldview. Maslow (1971) notes people ask how good a society does human nature permit, but he replies: 'how good a human nature does society permit?' People tend to be more altruistic than the economic model predicts (Brondizio et al. 2010). Emotions such as compassion, empathy, love and altruism are key components of the human behaviour repertoire (Manner and Gowdy 2010). In reality, human nature is as much based on *cooperation and sharing* as on greed. It depends on the society which of these comes to the fore.

Many societies have lived happily and sustainably for thousands of year precisely because they didn't pretend they were 'Masters of the Universe' who were in control of everything. So the key question to ask in Western society is 'Are we in control?' The environmental crisis clearly shows that as a society we don't know what we are doing; it is out of control. Hulme (2009) suggests that a confident belief in the human ability to control nature is a dominant, if often subliminal, attribute of the international diplomacy that engages climate change. In fact the confident belief we can control everything is embedded right through Western culture. Anthropocentrism encourages us to think we are the truly privileged 'top dog', the 'Masters of Nature'. Some seem to think that to stay at the top we must brutally overpower our enemy – being all other forms of life (Washington 2013b). Human mastery seeks to subjugate this unknown and scary natural 'other'. Under anthropocentrism, we are not loving brothers and sisters to the rest of life, but cruel masters seeking total domination. To *Homo economicus* (Daly and Cobb 1994), nothing is sacred, and beauty is just a thing of private preference. This is part of the psychology of anthropocentrism (involving paranoia, fear and denial) we discussed in Chapter 2.

It has been pointed out that it is better to build our cultures in intelligent harmony with the way the world *is,* rather than try to 'take control' as Western society has been doing (Rolston 2012). Rolston asks if we want nature to end. He wonders (p. 46):

> What kind of planet do we want? What kind of planet can we get? Maybe we also ought to ask: What kind of planet do we have? What kind of planet ought we to want? Maybe we ought to develop our capacities for gratitude, wonder, respect and restraint. Maybe we live on a wonderland Earth that we ought to celebrate as much as to develop.

The reality is we do not 'control' the Earth any more than we own it. Control implies knowing what you are doing, and clearly society doesn't know what it is doing – hence the environmental crisis leading us towards collapse (ecosystemic and societal). However, if we acknowledge this, accept we are part of nature *not* its master, then in so doing we overcome a major obstacle to healing our world.

'Land rights' in balance with 'obligation *to* the land'

We should also discuss a complex issue in regard to the issue of 'land rights' and how they relate to 'rights of the land' and to our obligation *to* the land. This can become an example of a 'conflict of rights', one of which is based on social justice, the other on ecojustice (Washington et al. 2018). As an Australian, I am very aware of the 'land rights' campaigns where Aboriginal people have sought to gain recognition of their traditional bond with the land. Indeed, I have supported such campaigns and worked with Aboriginal people over many years in regard to nature conservation. I also support Aboriginal people in their campaign to gain a Treaty in Australia (which is the only nation not to have such a treaty!). I also especially respect (and relate to) Traditional Aboriginal Law.

Graham and Maloney (2019) note that Aboriginal Law has two basic precepts: *the land is the source of the law;* and *you are not alone.* They note that in Aboriginal law, the land is a sacred entity, not property or real estate; it is the great mother of all humanity and is the source of life and law. They go on to state that because land is sacred and must be looked after, the relationship between people and land becomes the template for society and social relations. Therefore they say all meaning comes from the land or 'country'. Graham and Maloney (2019) refer to the 'law of obligation':

> As the land created us, so we are always going to be obligated to it. Not just our life but our *existence*, the whole of our existence and all meaning that underpins and surrounds it, that lives through us. All the flora and fauna, every living thing, all the landforms and features of the land, they are all our ancestors, because they all came before us. They helped us emerge and helped us to become human and to stay human, to develop us further as human beings and to create culture.

They also go on to note: 'Because we are always obliged to the land, always thankful for it, we are in turn obliged to look after it. This produces reciprocity – the land looks after us, we look after it' (Ibid.). In relation to the law 'You are not alone', they note:

> Aboriginal people maintain that humans are not alone. They are *connected and made* by way of relationships with a wide range of beings, and it is thus of prime importance to maintain and strengthen these relationships.

They further state that there are four essential attributes to the 'relationist ethos' of Aboriginal people (that are similar to those listed by Rose 1992) – *a custodial ethic, locality, autonomy* and *balance*. They observe that the custodial ethic is looking after the land, and note: 'Ethics only come from having empathy and from looking after something outside ourselves' (Graham and Maloney 2019). Locality refers to peoples' connection not just to country or nature generally, but to the region they come from, the particularity of their land. They note regarding balance that Aboriginal people: 'sought balance in everything: between people and the land; between men and women, with men's law and women's law; between different families and clan groups' (Ibid.). Similarly, in regard to Aboriginal Law and the 'Fitzroy River Declaration' in northwest Australia, Lim et al. (2017: 18) state that since the beginning of time: 'these First Laws have ensured the health of the living system of the Mardoowarra'. They state that: 'the river gives life and itself has the right to life' (Ibid.). This is in line with Maori views in New Zealand about rivers being living beings (Strang 2019).

In regard to Indigenous peoples in other parts of the world, Knudtson and Suzuki (1992: 40) observe in regard to Indigenous law that: 'Its ancient Laws remain timeless, eternally binding human beings to live in harmony with and respect for other species'. Knudtson and Suzuki (1992: 14) point out that:

> The Native Mind tends to view wisdom and environmental ethics as discernible in the very structure and organisation of the natural world rather than as the lofty products of human reason far removed from nature … It tends to see the entire natural world as somehow alive and animated by a single, unifying life force, whatever its local Native name. It does not reduce the universe to progressively smaller conceptual bits and pieces.

Essentially, all Australian Aboriginal tribes believed in 'Caring for Country' (Graham and Maloney 2019). Caring for Country conceives of the land and its non-human inhabitants as deeply embedded in both the practical use of natural resources and the spiritual nourishment of society (Head et al. 2005). It has been said re Caring for Country:

> This holistic approach emphasizes that not only are cultural, environmental, economic and societal systems not in opposition to one another, but they are necessary for each other's survival. It also has implications for notions of ownership and resource use. Caring for Country embodies an alternative view of ownership than those conceptualised by Western legal institutions. Ownership within Caring for Country represents a profound connection to place that goes hand in hand with stewardship relations of responsibility and obligation to Country, its human and non-human inhabitants.
>
> *(CoM/MSI 2016: 6)*

This is a very different idea of 'ownership' from that of Western society. In fact, *so different* is this idea philosophically that it is far better described as a connection, obligation to and 'duty of care' to nature. The use of the word 'ownership' above thus confuses the issue, for what is meant is totally at odds to the Western idea of land philosophically owned as a slave.

I am deeply aware of the deep injustices Aboriginal people have suffered due to colonisation (I was on the Board of Management of Mutawintji NP in Western NSW and heard many reports first hand from my Paakintji friends and colleagues). However, I am equally aware of the history of injustices heaped upon the land by Western colonisation. Hence I think we have to unpick the ethics involved in a complex issue of what in the West we call 'rights'. As noted before, many Indigenous peoples never believed they *owned* the land philosophically, yet today many are nevertheless caught up in battles over 'land rights' (though in Australia 'Native Title' is not legal title but recognition of rights and interests to land). However, the idea of 'rights' is itself a Western concept. Graham and Maloney (2019) state that in Australia: 'rights based laws and the notion of any living thing having a "right" is not a conceptual element in the Aboriginal legal system'. They use the term basically because 'whitefellas' in Australia use it; however, it is not from Aboriginal culture. They go on to note: 'Aboriginal law and ethics are based on relationships and based on a perpetual obligation to look after land and each other'. So 'obligation' to the land is central, rather than a belief that one has the 'right' to do whatever one likes.

However, the complexity of mixed ethics around the idea of 'ownership' is clear in part of the wording of the Uluru Statement in Australia (www.1voiceu luru.org/the-statement/) which says:

> This sovereignty is a spiritual notion: the ancestral tie between the land, or 'mother nature', and the Aboriginal and Torres Strait Islander peoples who were born therefrom, remain attached thereto, and must one day return thither to be united with our ancestors. This link is the basis of the ownership of the soil, or better, of sovereignty.

On the one hand sovereignty is said to be a 'spiritual notion' and a link to 'mother nature', but in the next sentence it says: 'this link is the basis of the ownership of the soil'. This brings together two ideas actually in philosophical and ethical conflict.

Of course, many lands where Indigenous peoples lived were invaded and colonised (causing great suffering to First Peoples *and* the land). Historically, Indigenous Australians were living in place quite sustainably, and were pushed off the land by invasion, often brutally. They thus deserve *restorative justice* (as part of social justice) to make amends for being pushed off the country they had a responsibility for, and an obligation to. However, in terms of ecological ethics and ecocentrism, that in no way means they ever 'owned' the land *philosophically* in the Western sense. The land owns itself (so that it is no person's slave) and people should have a role of custodianship and caring to look after country. That is a duty of care, however, not a deed of philosophical ownership.

However, Aboriginal people live inside an Australian society dominated by the Western idea of philosophical 'ownership' (and other anthropocentric ideology). Hence, inevitably some of them will accept that idea and claim Western 'ownership' of their ancestral lands. In many cases claiming 'ownership' may actually be undertaken simply to try to stop them being mined, logged or otherwise degraded by Western society. However, 'land rights' in much discussion has been used as a substitute for the restorative justice due to Aboriginal people. The trouble is the two are *not* the same, and ethically we don't provide restorative justice to Indigenous people through the idea of an 'ownership' that sees the land as a slave. However, in many cases this is all that 'land rights' will offer Indigenous people (thinking that this alone is justice).

So philosophical ownership of land (assuming it is a slave to people) is clearly at odds with the traditional Aboriginal belief of 'Caring for Country' as part of Aboriginal Traditional Law. If we actually care for the land, we can seek to *work with* the land in respect and reciprocity, which is a very different thing from seeing the land as a slave you own. There is also an odd idea in academia that influencing the land (e.g. by a fire regime) means you somehow 'create' it (see Washington 2018b for discussion). However, people didn't physically shape the land over millions of years to make mountains or valleys, neither did they evolve its species (most go back in time long before there *were* humans). First Peoples certainly *influenced* biological communities (primarily by fire regimes), but having influence in no way means that the land is a 'human artefact', as has been claimed (Gomez-Pampa and Kaus 1992), or a purely 'cultural landscape' (Langton 1996). These claims are mistaken on two levels. Firstly, it is wrong because it overstates the biophysical influence of First Peoples on the natural world. Secondly, the claim that humans 'create' the land is strongly anthropocentric, and ignores the intrinsic value, agency and rights of the land itself (Washington 2018b). This claim is fairly common, however, and has been taken to such an extreme that one writer (Gammage 2011) has claimed that Australia was just a 'great estate' managed and owned by Aboriginal people. However, rather than being a praiseworthy idea, in fact it assumes Aborigines believed in the Western anthropocentric idea of philosophical 'ownership' (when they did not). This again comes from a deeply anthropocentric worldview, very different historically from that of most Indigenous worldviews (Neidjie et al. 1985; Knudtson and Suzuki 1992; Curry 2011). In contradiction to that idea, Aborigines actually worked *with* the land and 'cared for country' (Graham and Maloney 2019). An Elder of Kakadu country in the Northern Territory, Big Bill Neidjie explains this (Neidjie et al. 1985: 46):

> This ground and this earth … like brother and mother. Trees and eagle … you know eagle? He can listen. Eagle our brother, like dingo our brother. We like this earth to stay, because he was staying for ever and ever. We don't want to lose him. We say 'Sacred, leave him' … our story

is in the land … it is written in those sacred places. My children will look after those places, that's the law.

Hence there is a lot of confusion about land rights, sovereignty, ownership and justice. Indigenous people in many nations clearly deserve restorative justice, and that may involve the *legal fiction* of ownership of land, but it should never involve the philosophical idea of ownership, which is so very much at odds with the kinship ethics common in Indigenous societies (Knudtson and Suzuki 1992).

Graham and Maloney (2019) argue that in general the idea of 'Rights of the Land' proposed by Earth jurisprudence (Cullinan 2003) conforms well with Aboriginal Traditional Law, apart from the fact that the idea of 'rights' is *not their term*, as they speak of 'obligations' to the land. However, land rights campaigns may conflict with 'Rights of the Land', which argue that nature should be given legal rights to exist. This highlights yet again the issue we found with population – that 'rights' are not absolute, they are relative and interact, and must be considered together. I argue here that the key 'right' of the land is *existence*. The land has a right to exist and continue in a healthy natural condition, as does the life that evolved there. Another way of saying this is that the premise of the ancient Aboriginal Law must operate – that we must consider first the health of the land (Lim et al. 2017; Graham and Maloney 2019). That right to exist (or Law that the health of the land comes first) should *supersede any other 'rights'*, so that in legal terms one could say the health of the land was 'paramount'. This is similar to the idea that life is the highest value of all, as expressed in 'the will to live' (Sweitzer 1949; Schopenhauer 1983). Keeping country healthy should thus be the key value in ecological ethics (Curry 2011; Rolston 2012; Washington 2018b) and is also what many Indigenous people say is their key ethic (Knudtson and Suzuki 1992; Graham and Maloney 2019).

BOX 8.1 EARTH JURISPRUDENCE, MICHELLE MALONEY

The term 'Earth jurisprudence' was coined by deep ecologist and 'geologian' Thomas Berry. In his book *The Great Work: Our Way into the Future*, Berry critiques the underpinning governance structures of modern industrial society and proposes that the fundamental human establishments that control human affairs – law and government, economics, education and religion – are all anthropocentric and have brought us to the ecological crisis we face today.

Berry calls for a new jurisprudence (philosophy of law), one that puts the community of life at the centre – as a foundation for human governance and an inspiration and guide for human activities.

Earth jurisprudence calls on us to situate ourselves within the interconnected community of life, rather than as separate from it. Recognition of human interconnectedness with nature is a prerequisite for ecological sustainability and should be recognised as the foundation of our legal system. Historically, the environmental conditions of civilisations have been altered by geographical, cultural and socio-economic changes. However, contemporary industrial civilisation has altered its environment to such a degree that its own existence has been placed in jeopardy.

The current legal system is human-centred and has been shaped to promote infinite economic growth and industrial development. This orientation effectively treats the environment as a resource that belongs to human beings and can be exploited for human benefit. Environmental law and other regulatory measures play an important role in curving the excesses of this system. However, the current rate of environmental destruction demands a new approach to environmental protection.

To this end, Earth Jurisprudence seeks to catalyse a shift in law from human-centredness to a recognition of our interconnectedness with nature. It promotes a number of key principles including: the intrinsic right of nature to exist and flourish, the need to create governance structures that enable human societies to fit within our ecological limits and the benefits of engaging with culturally diverse governance structures, including indigenous knowledge systems.

Earth Jurisprudence engages with a broad subject matter that includes (but is not limited to) legal theory, environmental law, economics, indigenous epistemologies, cosmology, science, religion and environmental philosophy. Innovative and creative work needs to occur in each of these areas if we are to build a truly sustainable and mutually enhancing future society. Today, 'Earth Laws' are emerging as a new constellation of ideas and principles, embedded in Earth jurisprudence and with specific expression in concepts such as: Rights of Nature, Ecocide and Governance for living within our ecological limits.

The Australian Earth Laws Alliance invites everyone to engage with a positive vision for the future of human governance and to play our part in transforming our relationship with the nonhuman world.

For further information: See: www.earthlaws.org.au/ and https://theright sofnature.org/

There are practical and ethical reasons for why I delve into this relation of 'land rights'/'rights of the land'/and obligation *to* the land. That is because Indigenous people everywhere are increasingly being pushed by corporations and governments (and the media) to *develop* (mine, log, graze, etc.) their land, even if this breaks Traditional Law (or lore). This of course is happening due to the endless growth obsession of Western society. A case in point is what happened at James Price Point in northwest Australia. Mathews (2013) reported on the proposal to establish a natural gas plant by Woodside Petroleum. This would have required dredging a channel, laying ocean pipes, constructing a major breakwater, as well as clearing 2,400 hectares of woodland. It would have required extensive blasting of coral reefs, caused sediment and sea pollution from drilling and dredging, seismic pollution dangerous to whales, and ongoing heavy shipping traffic (Ibid.). There were two Aboriginal groups involved who may have had 'Native Title'. One group (the Goolarabooloo) refused development, while the other group (the Jabirr Jabirr) accepted it (and the economic benefits it *might* bring them). Eventually, native title was given to the Jabirr Jabirr. So the development was given the green light to go ahead, and it was only because the developer determined the proposal to be uneconomic that it did not proceed. However, the Jabirr Jabirr have stated that in future they would welcome development opportunities like those that had been proposed by Woodside Petroleum (Aikman 2017).

Degradation of the land (and its life) breaches ecological ethics as well as Aboriginal First Laws. As Aboriginal Law states, people have an obligation *to* the land to keep it healthy. However, if we are to use the Western idea of 'rights', then I suggest a 'hierarchy of rights' to protect life and land. The land owns itself and has a right to a healthy existence – and that is the fundamental overarching 'right'. Human groups (especially Indigenous people) may have other rights (as well as obligations) in regard to land that should be respected. However, the 'rights of the land' must *always* come first before any human land rights. After all, for humans to be able to live in place on the land is a privilege which implies we have responsibility, obligation and a 'duty of care' to that land, based on respect and reciprocity, but never philosophical ownership (Washington 2018b). The land is paramount and should never be degraded by any group, even if they hold legal 'land rights'. A more fundamental ethic and right is involved – to care for country, to protect life.

Now I suspect that some people may be affronted by my above conclusion. I can only say that those who attack this conclusion are clearly putting social justice (restorative justice) *far above* ecojustice for nature. I do not believe that this approach is ethically correct, for to me both social justice and ecojustice are entwined (Kopnina and Washington 2019) and must operate together. Given the extent of the environmental crisis underway, it is time I believe for a 'Nature First' assumption in regard to planning (Washington and Kopnina 2019). This should assume nobody 'owns' the land philosophically, that it owns itself and that the rights of the land must come before the land rights of any human group. If we do not adopt such a position, then the environmental crisis will only continue to worsen. Protecting land and life is thus the fundamental right that we healers should uphold.

Conclusion

This chapter seeks to overturn a fundamental assumption in Western society – that humans *own* the land philosophically. Few of us ever stop to consider and ask: 'Do we really?' I just looked out the window on my land next to

Wollemi National Park and saw a Red-neck Wallaby hiding from the hot sun and a heat wave under a Brittle Gum. Earlier I heard the call of a Gang Gang Cockatoo as she flew past, and later I was lucky to see three of the even rarer Glossy Black Cockatoo! Supposedly, this is 'my' land, but *is it not theirs as well?* When I walk to my cliff and stare out over 50 km of forested ranges stretching into the distance, or go to the 'Happy Place' where the Wiradjuri people used to camp, how can I consider that such powerful places are slaves to *my* will? If anything they are wiser and smarter than I will ever be. Indeed, my writings come from listening to that wisdom (Washington 2018b). If we are to heal the world, then that means overturning a dominant and insidious ideology of ownership and mastery that has placed the world in bondage. The land is not our slave, neither are we hers. Rather, the land and its life are our kin, our home that we should celebrate in joy and respect, and have a duty of care to heal and protect.

Healing the land means dropping the rather odd (and supremely egotistical) idea that we are the 'Masters of the Universe'. We are not masters, we may be special, but that specialness comes from acknowledging our roots and our responsibilities to all life. All of life is in fact special, and healing the world means first and foremost upholding the intrinsic value of life. That doesn't mean we cannot make use of the land. After all, Indigenous peoples did this sustainably for tens of thousands of years, but they did it from a basis of respect and reciprocity (Abram 1996, 2010; Graham and Maloney 2019). It should be clear, however, that massive projects that destroy or degrade large areas of nature are never ecologically sustainable, nor in accord with healing the world. We should uphold Indigenous First Laws, along with the 'Nature First' principle. That means we are ethically entitled as healers to argue degradation of the land is *always wrong* (no matter who suggests it) and we should oppose such decisions. Healing the world I believe means ranking the 'rights of the land' to healthy ongoing existence in front of all human 'rights'. It is essential that Western society rediscovers what Aboriginal Law tells us – we have an obligation to protect life.

What can *I* do?

There are several things you can do to break free from the blueprint of ownership:

1)   Reject the 'blueprint' ideology that grids the world into blocks that humans philosophically 'own' and where the land is just a human slave. Rather, help free the land from the bondage of 'ownership' (Crist 2019).
2)   Live in place as a custodian, steward and carer. Certainly you may need the legal fiction of 'owning' a piece of land where you live. However, don't let that fiction mean you slip into the philosophical idea that you 'own' the land (and are its master).

3) Cultivate humility and humbleness in regard to the land, where you listen to the land and spend time feeling if any decision to alter the land is right or wrong. This means stepping back from the usual mad rush of life to consider the land's point of view. As Leopold (1949: 262) spoke of in his 'Land Ethic', we should:

> Examine each question in terms of what is ethically and aesthetically right, as well as what is economically expedient. A thing is right when it tends to preserve the integrity, stability, and beauty of the biotic community. It is wrong when it tends otherwise.

4) Uphold our obligation to, and our 'duty of care' to protect, the land. Uphold the 'right of the land' to healthy ongoing existence as the key right that must be upheld, along with human 'land rights'. That means supporting a 'Nature First' assumption in regard to planning (Washington and Kopnina 2019) and maintaining we have a *prime obligation* to protect life.

5) If you legally buy land, you may wish (in the spirit of Chapter 7) to create a nature ritual that makes clear you are actually a custodian and carer for this land, who will listen to it and operate from an obligation towards it.

References

Abram, D. (1996) *The Spell of the Sensuous*, New York: Vintage Books.

Abram, D. (2010) *Becoming Animal: An Earthly Cosmology*, New York: Vintage Books.

Aikman, A. (2017) 'Kimberley gas protest fought on false native title claim', *The Australian*, 29 November, 2017. see: www.theaustralian.com.au/national-affairs/indigenous/kimber ley-gas-protest fought-on-false-native-title-claim/news-story/e25d8071d0ae90272330a226 d04814ad. (accessed 28 February 2019).

Berry, W. (1976) *The Unsettling of America*, San Francisco: Sierra Club Books.

Brondizio, E., Gatzweiler, F., Zografos, C., and Kumar, M. (2010) 'The socio-cultural context of ecosystem and biodiversity valuation', in *The Economics of Ecosystems and Biodiversity: Ecological and Economic Foundations*, ed. P. Kumar, London: Earthscan, chapter 4.

CoM/MSI. (2016) 'Caring for country: An urban application: The possibilities for Melbourne', City of Melbourne/Melbourne Sustainability Institute, see: www.monash.edu/__data/ assets/pdf_file/0006/720681/Caring-for-Country-Literature- Review.pdf. (accessed 4 March 2019).

Crist, E. (2019) 'Let earth rebound! Conservation's new imperative', in *Conservation: Integrating Social and Ecological Justice*, eds. H. Kopnina and H. Washington, New York: Springer, pp. 201–218.

Cullinan, C. (2003) *Wild Law: A Manifesto for Earth Justice*, Totnes, Devon: Green Books.

Curry, P. (2011) *Ecological Ethics: An Introduction*, 2nd ed. Cambridge: Polity Press.

Daly, H. and Cobb, J.. (1994) *For the Common Good: Redirecting the Economy toward Community, the Environment, and a Sustainable Future*, Boston: Beacon Press.

Gammage, W. (2011) *The Biggest Estate on Earth*, Sydney: Allen and Unwin.

Gomez-Pampa, A. and Kaus, A.. (1992) 'Taming the wilderness myth', *Bioscience*, **42** (4): 271–279.

Graham, M. and Maloney, M.. (2019) 'Caring for country and rights of nature in Australia – A conversation between earth jurisprudence and aboriginal law and ethics', in *Sustainability and the Rights of Nature in Practice*, eds. C. La Follette and C. Maser, (forthcoming 2019), Florida: CRC Press.

Harvey, G. (2005) *Animism: Respecting the Living World*, New York: Columbia University Press.

Head, L., Trigger, D., Mulcock, J., et al. (2005) 'Culture as concept and influence in environmental research and management', *Conservation and Society*, **3** (2): 251.

Hulme, M. (2009) *Why We Disagree about Climate Change: Understanding Controversy, Inaction and Opportunity*, Cambridge: Cambridge University Press.

Josephy, A. M., Jr. (1995) *500 Nations: An Illustrated History of North American Indians*, New York: Hutchinson/Pimlico.

Knudtson, P. and Suzuki, D.. (1992) *Wisdom of the Elders*, Australia: Allen and Unwin.

Kopnina, H. and Washington, H. (2019) *Conservation: Integrating Social and Ecological Justice*, New York: Springer.

Langton, M. (1996) 'The European construction of wilderness', *Wilderness News (TWS)*, summer 95/96.

Leopold, A. (1949) *A Sand County Almanac, with Essays on Conservation from Round River*, New York: Random House (1970 printing).

Lim, M., Poelina, A., and Bagnall, D. (2017) 'Can the Fitzroy river declaration ensure the realization of the first laws of the river and secure sustainable and equitable futures for the West Kimberley?' *Australian Environment Review*, **32** (1): 18–24.

Manner, M. and J. Gowdy. (2010) 'The evolution of social and moral behaviour: evolutionary insights for public policy', *Ecological Economics*, **69** (4): 753–761.

Maslow, A. (1971) *The Farthest Reaches of Human Nature*, New York: Viking Press.

Mathews, F. (2013) 'Why is the Kimberley calling?' see: www.kimberleycalling.org/why-is-the-kimberley-calling.html. (accessed 4 March 2019).

Neidjie, B., S. Davis, and A. Fox. (1985) *Kakadu Man: Bill Neidjie*, Australia: Mybrood P/L Ind (Allan Fox and Associates).

Neihardt, J. (1972) *Black Elk Speaks: Being the Life History of a Holy Man of the Ogalala Sioux*, New York: State University Press of New York.

Rolston, H. III. (2012) *A New Environmental Ethics: The Next Millennium for Life on Earth*, New York: Routledge.

Rose, D. B. (1992) *Dingo Makes Us Human*, Cambridge: Cambridge University Press.

Schopenhauer, A. (1983) *The Will to Live: The Selected Writings of Arthur Schopenhauer*, ed. R. Taylor, New York: Ungar.

Standing Bear, L. (1928) *My People the Sioux*, New York: Houghton Mifflin.

Standing Bear, L. (1978) *Land of the Spotted Eagle*, Nebraska: University of Nebraska Press.

Strang, V. (2019) 'The rights of the river: Water, culture and ecological justice', in *Conservation: Integrating Social and Ecological Justice*, eds. H. Kopnina and H. Washington, New York: Springer, pp. 105–120.

Sweitzer, A. (1949) *The Philosophy of Civilisation*, New York: Macmillan.

Tennyson, A. (1849) 'In Memoriam A.H.H.', poem by Alfred, Lord Tennyson, in *Alfred Lord Tennyson, Poems*, ed. H. Tennyson (1890), London: Macmillan.

Vetlesen, A. (2015) *The Denial of Nature: Environmental Philosophy in the Era of Global Capitalism*, London: Routledge.

Washington, H. (2013a) *Poems from the Centre of the World*, lulu.com, see: www.lulu.com/shop/haydn-washington/poems-from-the-centre-of-the-world/paperback/product-21255751.html. (accessed 27 March 2019).

Washington, H. (2013b) *Human Dependence on Nature: How to Help Solve the Environmental Crisis*, London: Earthscan.

Washington, H. (2018a) 'The intrinsic value of geodiversity', *The Ecological Citizen*, **1** (2): 2. see: epub-007, see:, www.ecologicalcitizen.net/article.php?t=intrinsic-value-geodiversity.

Washington, H. (2018b) *A Sense of Wonder Towards Nature: Healing the Planet through Belonging*, London: Routledge.

Washington, H., Chapron, G., Kopnina, H., Curry, P., Gray, J., and Piccolo, J.. (2018) 'Foregrounding ecojustice in conservation', *Biological Conservation*, **228**: 367–374.

Washington, H. and Kopnina, H.. (2019) 'Conclusion: A just world for life?' in *Conservation: Integrating Social and Ecological Justice*, eds. H. Kopnina and H. Washington, New York: Springer, pp. 219–228.

9
LISTEN AND WONDER

Sometimes we are gifted
With a special moment,
A day of meaning
Where all things
Come together
And one smiles
In sheer wonder.
With a sudden silence
A caress of wind
And a brief, ethereal
Ripple of light -
It is as if
The Goddess passes
And bestows a smile
Upon the open heart.

From 'The Gift' (Washington 2013a)

SUMMARY

This chapter discusses something central to healing the world – our *sense of wonder* towards nature. All children feel this, but it often gets buried in adults (for reasons discussed). The sense of wonder towards nature is present in the writing of many cultures and expressed (in various words) by many scholars. In fact, its importance has clearly been rediscovered again and again throughout history. One reason why wonder is often buried in Western culture is its anti-spirituality, where it actively opposes a spiritual connection to nature. Part of the reason is the deeply embedded anthropocentrism in Western society. Another is that those spiritually connected to nature do not make good consumers who 'Shop till they drop!' The chapter proposes the reawakening of an Earth spirituality and how we can re-enchant the world. The chapter discusses why rejuvenating wonder is a key part of healing the world, as it gives us the *deep belief* to act. The chapter concludes by listing the steps we can take on the road back to wonder. For the detail explaining this summary, please read on; however, if you just want to know the things *you can do* (six are listed) about this, then turn to the end of the chapter.

Introduction

Many of us (at least at times) feel a *sense of wonder* towards nature. Many of us can also remember this wonder from our childhood. A yearning for a 'sense of place' is a perennial human longing. All peoples need a sense of 'my country', of belonging to a sustaining landscape they respond to in care and love (Rolston 2012). Wild nature can be valued as a place which restores one's 'sense of wonder' in life (Washington 2018a). This can also be called a 're-enchantment' of the land (Tacey 2000). Part of this sense of wonder is a feeling of being at 'one' with the land, of *belonging* (Thomashow 1996). This issue fascinated me to the extent I self-published a small book on wonder (Washington 2002) and greatly expanded this for my recent book *A Sense of Wonder Towards Nature* (Washington 2018a). So, what is this sense of wonder towards the natural world? Is it something innate we are born with, or can anybody learn it? Is it something more common in different races, or merely different societies? Does the fact that only 20% or so of people (in Western society) get interested in environmental issues mean that only that percentage can feel the sense of wonder? Alternatively, does the growing interest in the environment (especially climate change) in young people mean that the number who can feel a sense of wonder is growing? These are important questions to consider.

It is common for environmentalists to list all sorts of scientific arguments about why a natural area should be saved. It should be saved because it is a reservoir of biodiversity. It should be saved because there are 'X' threatened species. It should be saved because it has unique geodiversity, because it provides ecosystem services to society. It should be saved because it provides clean air and clean water to society. I know this list well, as I have myself run campaigns to protect wild places using similar arguments. These arguments are of course valid, and this book lists similar reasons for valuing nature. However, it seems that through our history of scientific analysis we have lost the ability to see nature as an 'interlocking whole' (Collins 2010). Infrequently do we argue that a natural area should be saved because it is a place of 'wonder' we love, or because it is 'sacred'. To say this in Western society would mean that one's words could then be dismissed as 'emotional'. It is strange that we have come to such a pass, if we cannot get emotional about the land we love – what *can* we get emotional about? Or is our passion dead?

The sense of wonder I am talking about here is a deep connection with the land. As such it must be considered 'spiritual', even if one does not call it religious. An Agnostic, a Christian, a Buddhist, a Hindu, a Muslim … all can feel a sense of wonder, even if they call it by varying names (Washington 2018a). What I am talking about here is not something that should be trivialised and put on the shelf with the books on fairies. I am talking about the fundamental relationship of humanity with the land, a bond which has nurtured us for as long as our species has existed. I am talking about one of the deepest and most abiding loves of them all: the love of the land. There are also high points in our sense of wonder. I have called them 'illuminating moments' (Washington 2018a) while others have called them 'transcendent' (Louv 2005) and a 'hierophany' or epiphany (Oelschlaeger 1991). Such moments can change one's life. Wilderness advocate John Muir was escaping from the military draft in 1864 and was near Lake Huron in Canada. He wandered into a dark swamp where he came upon a cluster of rare white orchids (*Calypso borealis*). They were so beautiful that he later wrote (in Corcoran 2001):

> I never before saw a plant so full of life; so perfectly spiritual. It seemed pure enough for the throne of its Creator. I felt as if I were in the presence of superior beings who loved me and beckoned me to come. I sat down beside them and wept for joy.

The sense of wonder is also about *listening* and 'empathy'. One must let down one's guard, open oneself up and let all one's senses absorb the beauty of the natural world. Aboriginal Elder Miriam-Rose Ungunmerr (see Tacey 2000) explains that there is a word in the Ngangikurungkurr language 'dadirri', which is: 'something like what you (white people) call "contemplation"'. Tacey (2000) believes this is a spirituality of deep seeing and deep listening. This would seem to me to be part of the empathy that allows us to feel the sense of wonder. Ungunmerr believes that the gift of dadirri is perhaps the greatest gift that Aboriginal people can give to their fellow Australians (Tacey 2000). The writings of other scholars, and my own lived experience, lead me to understand that a sense of wonder is an essential part of the human psyche (Washington 2018a). It could play a key role in helping us reach true sustainability. It can also help to solve 'Nature Deficit Disorder' (Louv 2005). Healing the world will require that we use our emotional intelligence as much as our intellectual intelligence (Hempel 2014). Our sense of wonder at nature is thus not just 'one more value', but something of central importance, something that gives us *meaning* (Washington 2018a).

A truth endlessly rediscovered ...

I have read hundreds of books about the environment over more than four decades. Early on I found key classics such as Thoreau's (1854) *Walden*, Leopold's (1949) *A Sand County Almanac* and Berry's (1988) more recent classic *The Dream of the Earth* and Snyder's (1990) *The Practice of the Wild*. Since then I have found several other key works (note I select just a few here) such as Carson's (1965) *The Sense of Wonder*, Oelschlaeger's (1991) *The Idea of Wilderness* and Abram's wondrous books *The Spell of the Sensuous* (1996) and *Becoming Animal* (2010). I was also reading David Suzuki's books, such as Knudtson and Suzuki (1992) *The Wisdom of the Elders*. In the last few years I have been privileged to find Curry's (2011) *Ecological Ethics*, Rolston's (2012) *A New Environmental Ethics*, Vetlesen's (2015) *The Denial of Nature* and Weber's (2016) *The Biology of Wonder*. This is without mentioning many other books discovered along the way that relate to ecocentrism, wilderness and sustainability (as discussed in Washington 2015, 2018a).

I also found that other cultures had recorded their sense of wonder, going back millennia (Washington 2018a). I came to understand that Indigenous cultures (Australia's own First Peoples being a highlight) had (and still have) a deep, ecocentric and kinship-based ethics (e.g. Neidjie et al. 1985). Now looking at these cultures, I could see that they found the same truth again and again in different parts of the world – that the rest of life was our kin and deserved respect. Similarly, when I consider all of those books above, I realise some authors would have read a few other authors (especially the works of Thoreau, Muir and Leopold) but many would not have done so. Instead, they too had discovered (or perhaps rediscovered) the same truth. How? Because *if you listen to the land, you will learn*. I certainly have and continue to do so. People have been doing this throughout human history; indeed I believe it is the most obvious and important thing that humanity can do. It is through doing this that we learn wisdom and mature as both people and societies. Only in the last few centuries have we turned away from the 'Wisdom of the Elders' and pretended that only humans were truly 'alive' and had meaning, while the rest of nature was just a machine, mere dead matter-in-motion (Vetlesen 2015). The result has been the environmental crisis and the huge extinction event now underway (see Chapter 1). The scale of what society is doing overall can only be described as a great moral crime.

My point here, however, is to remind society that humanity has discovered the importance of ecocentrism and a sense of wonder again and again. Many environmental scholars continue to do so, even if they use a variety of terms to attempt to describe this. Indeed, it should be obvious to anyone with an open heart who has set foot in wild nature and heard a wild bird sing and a wild brook's splashing song (Washington 2018a). It is a great comfort to me to know there are fellow listeners, that there are many people out there who have listened to the land, felt the same sense of wonder and learned the same truth. So the truth is there to be found; we can find our way back to wonder and to wisdom as to how to live in balance as part of (not master of) this amazing living world.

Anti-spirituality in Western culture

Part of the cause of why our sense of wonder has declined is that modern Western society does not just ignore environmental ethics and a sense of wonder, it *actively opposes* a deep spiritual connection to nature. Tacey (2000: 185) argues:

> The consumerist society reinforces our shrunken empty status. It is vitally important for capitalism that we continue to experience ourselves as empty and small, since this provides us with the desire to expand and grow, and this desire is what consumerism is based on. Consumerism assumes that we are empty but permanently unable to fulfil our spiritual urge to expand. It steps into the vacuum and offers its own version of expansion and belonging … If we stopped believing in the myth of our shrunken identity, the monster of consumerism would die, because it would no longer be nourished by our unrealized spiritual urges. Therefore true spirituality … is extremely subversive of the status quo. This is why consumer society is keen to debunk or ridicule true spirituality.

Why do so few people today seem to feel a sense of wonder? Why is it relegated to children, and only survive in a minority of adults (at least in Western society)? Where did we go so wrong that our society acts to *suppress* a sense of wonder towards nature, rather than celebrate and foster it? This is the fundamental alienation of oneself from the natural world. Clearly something is very wrong for us to have come to this state. Clearly our whole worldview must be at fault (Berry 1988). How could we let our sense of wonder, that connection that gives life meaning, slip away? It is clear that nature in Western society is no longer sacred, no longer the 'Earth Mother' as it was in the 'old sustainability' (Washington 2015). It has become just a 'resource' for human use (Crist 2012). The underlying worldview of Western society is thus at great pains to suppress a sense of wonder and a spiritual connection with nature (see Chapter 2). This is something that desperately needs to change if we are to heal our world (Washington 2015, 2018a).

Reawakening an earth spirituality

The opportunity is there for us to rediscover the sense of wonder. Tacey (2000: 262) sees this as being not so much a new religious ideology as a: 'new religious attitude that respects the diversity of all spiritual expressions'. So change is underway, it just needs to be accelerated. Tacey (2000: 7) speaks of a possible future where:

We are about to experience what could be called a 'second' enlighten-ment, a postsecular enlightenment, where religion and spirituality will return to centre stage and where secular materialism will appear out of date and anachronistic.

He believes this awareness will avoid established religions in favour of a more direct relationship with the sacred. Many people today seem to shy away from words such as 'spirituality' or 'the sacred'. However, the sense of wonder can equally well be described as an experience of the sacred spirituality of the land. The words should not hold us back – it is the *experience* that matters. We are at a point where the minority stream of thought (ecocentrism) needs to flower and replace the current dominant mentality (anthropocentric and nature-conquering). Change is often not gradual and linear, but sometimes arrives like the floodwaters from a collapsed dam. Remember the unforeseen unravelling of the Soviet Union? We may be poised for a rapid shift away from anthropocentrism and a rediscovery that we are part of nature (which entails respect, reciprocity and responsibility). This is my fervent hope. It would be wondrous also if *you too* would be part of that change, by rediscovering or rejuvenating your own sense of wonder (Washington 2018a).

BOX 9.1 EARTH SPIRITUALITY, DAVID TACEY

Many recognise that the world's major problems have spiritual elements that are not adequately acknowledged, partly because we don't know how to con-duct public debate at that kind of level. The ecological problem is at bottom a spiritual problem, an issue about our lack of feeling for, and identification with, the world as part of our larger spiritual being. Here indigenous people are our guides, because they see the spiritual as integral to the whole of real-ity, and those of us who are non-indigenous are slowly being encouraged to do the same. To take indigenous wisdom seriously means overcoming some of our entrenched prejudices, including the Cartesian dualism that has been fundamental to how we view ourselves and the world.

The ecological emergency is forcing us to realise the detrimental effect of the dualism between humanity and nature that we have taken for granted. Instead of being part of nature we have seen ourselves as apart from it, lords or masters of creation. The objectification of nature, which has paralleled in some ways the objectification of women in patriarchal society, has brought a breakdown in the order of things. We have become all too familiar with the components of this breakdown: pollu-tion, depletion of the ozone layer, loss of biodiversity, climate change and an increase in extreme weather events. Something is not right, and many of us feel this in our bones, even if we are unable to express it in holistic terms.

Even people without much romantic feeling for nature are saying things like: 'Gaia is angry'. There are disruptions going on in the natural world that are unprecedented, and part of our response to the disorder is to think of nature as a unified whole, and resort to phrases such as the 'anger of Gaia' to account for it. This is a good thing that has arisen from the crisis: we are starting to think in holistic terms about nature and the crisis is reawakening in us the ancient mythological bond with the world. This bond has been lost, and although it is too early to talk about a spiritual renaissance, I believe the ecological emergency will force us in this direction.

I am in favour of green politics and environmental activism, and offer my support whenever I can. But they do not go far enough, and many in the environmental movement are aware of this. There is something missing, which we are too embarrassed to admit, because we are supposed to be postmodern and beyond this ancient mystery called 'spirituality'. We need a cosmology, a spiritual vision, to back up and support our green politics and laws. Secular dictates and policies are not enough to bring about the transformation that is required. Something has to be activated in the soul which is beyond the ego, with its habit of caring for itself. The encapsulated condition has to be overcome, so that we can respond to a universal extension of spirit into the world.

For more information

Tacey, D. (2004) *The Spirituality Revolution: The Emergence of Contemporary Spirituality*, East Sussex, UK: Brunner-Routledge.

Changing society's worldview will take a great deal of education. First of all we need to encourage our children to think for themselves, not mouth platitudes. We reach true education and 'deep learning' by encouraging young adults to think for themselves (Washington 2018a). We change denial by educating people about the facts, and showing them the positive solutions that exist (see Chapter 11), so they don't just close their minds because it is 'all too hard'. We change greed and 'looking after number one' by educating about ethics and altruism (Ibid.). We can redesign materialism and consumerism to create a sustainable civilisation (Assadourian 2013). We change mindless competition by educating about the cooperation present both in society and nature (Weber 2016). We educate our children about their 'oneness' and connection with nature. The importance of ecocentric education and education for wonder is thus critical (Washington 2018b).

So to return to the key concern, *how* do we bring back the sense of wonder in Western society? A key part of the solution lies in considering the perception of nature during childhood (as noted by Louv 2005, 2011 and others). Emerson (1836: 10) noted: 'The lover of nature is he whose inward and outward senses are still truly adjusted to each other; who has retained the spirit of infancy even into the era of manhood'. This observation remains as true today as it was back then. Evernden (1985: 112) points out that the child's first experience of nature might be marked by delight, fear or amazement, but that more importantly there is a: 'realisation that there is an other, something in experience which cannot be contained in the self and is, therefore, uncanny – and *wild*'. Merleau-Ponty (1962: xvi) has stressed the importance of 'being filled with wonder' at the world, so that when the child discovers otherness for the first time, it involves eliciting a perception of the world: 'upon which our idea of truth is forever based'. Edith Cobb (in Nabhan and Trimble 1994: 28) emphasises the potential of ages 7 to 12, when children are 'in love with the universe' and 'poised halfway between inner and outer worlds'.

Evernden (1992: 116) considers the effect anthropocentrism has on the imagination, arguing:

> Through our conceptual domestication of nature, we extinguish wild otherness even in the imagination. As a consequence, we are effectively alone, and must build our world solely of human artifacts. The more we come to dwell in an explained world, a world of uniformity and regularity, a world without the possibility of miracles, the less we are able to encounter anything but ourselves.

It is clearly essential for the child to have time to interact with wild nature (wild otherness) and accept it as an equal. Washington (2018a) discusses the experience of Annie Dillard (1992), where she surprised a weasel and for a moment she *became* the weasel. Having shared a similar moment with my lyrebird (Washington 2018a), I very much identify with her description. Evernden (1992) notes that humans have drained away wildness through supposedly 'wise management', and that native animals (for his region) such as the weasel are now uncommon. However, Evernden (1992: 22) suggests:

> But if there are not weasels, perhaps there will be something even smaller, less furtive, that a child-sized person could still surrender to. As long as there is hope of contact, some wild spot the gardeners and developers have missed, there is still the chance that some imagination may escape, untamed, into the green chaos.

We in the West urgently need to reconsider our view of nature, and Evernden believes the key lies in our imagination, which Baccelard (1969: 123) claimed: 'is not, as its etymology (name) suggests, the faculty of forming images of reality, it is the faculty for forming images which go beyond reality, which *sing* reality'.

What does all this mean? Nurturing the imagination of the child is thus not merely letting them have their world of make-believe; it is allowing them to bond with the 'otherness' of nature and sustain their sense of wonder. If we allow imagination to survive, if we let the child experience wild nature (untrammelled and undomesticated), if we let them think for themselves, and even more *feel* for themselves, then we allow them to mature their sense of wonder (Abram 2010). Perhaps in truth we just let them retain into adulthood the sense of wonder they already have – we let them belong in a living world full of mystery. These are not such hard things to do – but their importance is belied by their simplicity! I believe they are the basis of healing the environmental crisis and moving towards a truly ecologically sustainable future.

Speaking out for re-enchantment

So where does this take us on the road to healing the world? Clearly it is time for all those who feel a sense of wonder at the natural world to *speak out*. There is no reason to feel ashamed about having a sense of wonder towards nature, there is no reason to hide your sense of wonder because you fear it will be labelled as 'unscientific' (Washington 2018a). Rather, to feel a sense of wonder is a cause for pride and joy, for with the sense of wonder we can see the joy that is at the heart of all things. We need to speak out for a re-enchantment with nature (Tacey 2000). There should be no reason anymore for people to give me a second look when I tell them I am working on a book about the 'love of the land', or if I mention a place as being 'sacred' (Ibid.). Why should we apologise about being passionate about life, about nature, about feeling one with the natural world? Yet up to now, so many of us seem to have been just that – too embarrassed to admit we love the land and have a sense of wonder (Ibid.). This shows just how all-encompassing and effective has been the brainwashing of anthropocentrism and what Weber (2016) calls 'bio-economical thinking'.

Evernden (1992) suggests that if 'wild otherness' is to be sustained, then we need those who speak for the wild and the experience of bonding with nature to come forth and talk about it. Evernden (Ibid.: 133) concludes:

> It is speakers such as these that may help us acquire the vocabulary needed to accommodate wildness and extinguish the technological flashfire of planetary domestication.

It is clearly time for those of us who feel the sense of wonder to *speak out* and help others rejuvenate their own sense of wonder.

A path back to wonder

So how can we rediscover our sense of wonder and our love of the land? And further – is it important that we do so? To answer the second question first, it is absolutely and totally essential that humanity rediscover its sense of wonder. Why? Because without this it is very unlikely we will heal our world. A sense of wonder towards nature provides the deep belief we need to make hard decisions (Washington 2018a). Bringing back our sense of wonder is not just a peripheral issue, it is one of the *key steps in saving a world*. As David Tacey (2000: 167) has eloquently said, we must:

> [a]llow the sacred to flood back into the world, to become sensitive again to the mysterious presence or numinosity of the physical environment. We need to develop ... a postmodern animism, a heartfelt response to landscape, a tao of architecture, a zen of design, a mysticism of the ordinary. Our joy and love must be redirected towards the external world, by becoming receptive to the mystery of the world and by rediscovering our essential kinship with it.

Rediscovering a sense of wonder is also a key step in healing our sundered souls (Washington 2018a) that have been imprisoned in our own minds. A sense of wonder helps us find who we truly are (see Chapter 7). On all counts I believe that this is not just 'something worth doing'; it is *the* thing worth doing. So rediscovering our sense of wonder is something critical we must do in Western society. How then do we go about it?

First comes the question of anthropocentrism (see Chapter 2). Changing a worldview is not like changing your socks – it takes time. Evernden (1992) implies it is time for the emergence of the 'ultrahuman', the person who goes beyond the strictures of dualism. He argues (pp. 129–30):

> Perhaps there is only one conclusion a reflective naturalist can come to: that if we would protect nature from the perils of the 'environmental crisis', we must first acknowledge those perils arose as a consequence of conceptual imprisonment. If we would save the world, we must first set it free.

Indeed we must. David Suzuki (1998: 261) has this to say about our Western view of ourselves as being separate:

> We have to define the issue differently. The environment isn't something 'out there', something we take from, put into, and have to manage so we can continue to exploit it without damaging its productive capacity. The issue is far more profound. We literally are the Earth. There is no 'out there' separated from us by our skin.

It follows that while we may wake up and think 'I' am awake, and walk around in just one body, we are in fact connected to everything else around us, and the boundary of what we define as 'I' (which we generally in Western society have seen as our skin) is really quite arbitrary (Abram 2010; Fisher 2013). The sooner we realise this, the sooner we can leave dualism behind (along with the idea one is an 'ego in a bag of skin', Watts 1973).

So how do we rid our society of anthropocentrism? Well, there are moves towards this, and we need to accelerate this trend. However, many parts of academia remain stubbornly anthropocentric, whether they come from the Right or the Left (Kopnina et al. 2018). The key factor is to change our worldview and ethics (see Chapter 2). Books such as *Dark Green Religion* (Taylor 2010), *Becoming Animal* (Abram 2010), *Ecological Ethics* (Curry 2011), *A New Environmental Ethics* (Rolston 2012), *Radical Ecopsychology* (Fisher 2013), *The Biology of Wonder* (Weber 2016) and my own book *A Sense of Wonder Towards Nature* (Washington 2018a) are helping to catalyse a change in academia. I hope this book will also contribute to this. A new journal *The Ecological Citizen* (see www.ecologicalcitizen.net/) is leading the way academically through advocating ecocentrism. Indeed, I was involved with this journal in the creation of a 'Statement of Commitment to Ecocentrism' on their website (EC 2018). You too may wish to sign this? Now some people tend to say: 'But what use is it for me to speak up?' My response is that society actually mostly changes through small but meaningful steps (Hill 2001). Anthropologist Margaret Meade has noted that throughout history, small groups of determined people have changed society. She said: 'Never doubt that a small group of thoughtful, committed citizens can change the world. Indeed it is the only thing that ever has' (used with permission, InterculturalStudies 2010).

As the compassionate economist and philosopher John Stuart Mill (1857) noted, every great movement must experience three stages: ridicule, discussion, adoption. We are certainly in the second stage of 'discussion', perhaps moving towards adoption of ecocentrism. We need to do this quickly. An essential step along that path is the reawakening of our sense of wonder. Tacey (2000) believes that spirituality (as opposed to Western 'religion') is not just some relic from the ancient past, but a dimension of our experience that is contemporary, relevant and vitally important for mental health and social cohesion. Tacey (2000: 12) quotes research as showing that: 'an overwhelming majority of secondary school teachers … believe that spirituality is a vital ingredient in the education of youth'. He argues that some areas that lead the way to solutions are the experience of nature and landscape, the environmental emergency, engagement with Indigenous peoples, the visual arts, popular life-history and storytelling, biography and public interest in Eastern religions. All of us can be part of this wondrous transformation.

Key steps on the road to wonder

Washington (2018a) lists key steps to help us return to wonder in our society. We need to:

* *Belong.* That is, we need to 'arrive' and accept that the land in which we live is truly 'home'. As Snyder (1990: 40) has said:

  We must consciously fully accept and recognize that this is where we live and grasp the fact that our descendants will be here for millennia to come.

> Then we must honor this land's great antiquity – its wildness – learn it – defend it – and work to hand it on to the children (of all beings) of the future with its biodiversity and health intact. ... Home – deeply, spiritually – must be here.

Indeed, by belonging to the land we take a major step in healing the planet. Aldo Leopold (1949) wrote of his visionary 'Land Ethic', and we need to adopt (or relearn) such an ethic, one that makes our kinship with the land and its wonder a central tenet.

- Realise that the greatest gift we can give our children is that of learning to *think and feel for themselves*. This helps them to survive peer-group pressure and develop their own ideas. We can ensure that a forum exists within our education system (and all educational institutions from school on up) for a discussion of values. This discussion needs to cover anthropocentrism vs ecocentrism – nature as an ally (and brother/sister/mother) rather than a 'thing' or enemy.
- *Learn from Indigenous 'Law' and lore.* We need to examine science from the viewpoint of the Indigenous perspective, as Knudtson and Suzuki (1992) urge. Weber (2016) notes that the chill, abstract languages of the sciences place a barrier between us and the Aboriginal feeling of life. It is time for science to be taught in a holistic manner, and in a manner which takes in some of the aspects of Indigenous worldviews, and sees the parts and processes of the Universe as 'holy' and sacred. As Knudtson and Suzuki (1992) point out, in an Indigenous worldview, the integrity of nature might be described as a 'sacred ecology'. They point out (p. 180) that in science (as in Indigenous societies) genuine wisdom should be attributed to those:

> [W]ith the capacity to feel, to exhibit compassion and generosity toward others, and to develop intimate, insightful, and empathetic relationships not just with fellow human beings but, in some sense, with the entire membership of the natural world.

They argue for a 'sacred ecology' that unabashedly embraces and sanctifies nature, yet remains informed by the most subtle and compelling truths of modern science. Weber (2016) similarly calls for a new 'ecology of poetics'. Tacey (2000) points out that traditional Aboriginal people in Australia are happy to share their understanding of the land. That is my experience also.

- *Teach science in a creative, holistic way* and try to bridge the gap between science and art. We need to remember the limitations of reductionist science (Weber 2016). When asked if absolutely 'everything' can be expressed scientifically, Albert Einstein (in Clark 1971: 186) observed: 'Yes, it would be possible. But it would make no sense. It would be description without meaning – as if you described a Beethoven symphony as a variation in wave pressure'. It is time to insert meaning and ecological ethics into any teaching of science. I am certainly not the first to point this out. Indeed, in a statement entitled 'Preserving and Cherishing the Earth: An Appeal for Joint Commitment in Science and Religion' (EE 1990), notable scientists wrote:

> As scientists, many of us have had profound experience of awe and reverence before the universe. We understand that what is regarded as sacred is more likely to be treated with care and respect. Our planetary home should be so regarded. Efforts to safeguard and cherish the environment need to be infused with a vision of the sacred.

There is also a need to link science and art more effectively. As a final point on this topic, who more appropriate to quote than eminent environmental scientist Paul Ehrlich (1986), who has fought for decades to increase environmental awareness, both in science and throughout the community. He concluded (1986: 17):

> I am convinced that such a quasi-religious movement, one concerned with the need to change the values that now govern much of human activity, is essential to the persistence of our civilization.

I would like to emphasise that I am not demeaning the value of science. What I am arguing for here is a change in the way science is taught, and the introduction (or reintroduction) of creativity, ecocentric ethics and holism into science. Weber (2016) supports a new view of life – 'biosemiotics', where biosemioticians consider 'feeling and value' to be the foundation of all life processes.

- *Educate for wonder* (Washington 2018b). We need to prioritise education for wonder in the home, at school, at tertiary level and in the community in general. For schools, this means getting our students (especially primary) out into nature for experiential education to counteract 'Nature Deficit Disorder' (Louv 2005, 2011). It means changing the curriculum to foreground education for wonder (Washington 2018b).

- Teach *ecological economics* and abandon traditional neoclassical economics, which is based on an absurdity of growthism and denies any ecological limits. Indeed, the eight key assumptions of neoclassical economics are all irrational or unethical (see Chapter 5). We need a movement within economics which accepts ecological limits, believes in greater equality and encourages a worldview that is ecocentric. Luckily, this has already existed since the 1970s, being the *steady state economy* (Daly 1991, 2014; Daly and Cobb 1994; Dietz and O'Neill 2013).

- Cherish and nurture the *imagination* of our children, the importance of which is argued by many writers in regard to their connection to the natural world. If we can keep the imagination of our children alive into adulthood, then their chances of retaining a sense of wonder are greatly increased.

- Nurture the *creativity* of our children as well as getting them to learn facts and pass tests. Through self-expression and art we build a bridge to wonder, we free our Muse and empathic sense – so that we have a chance to truly listen and feel the sense of wonder.
- Give special support to children/adolescents over the period of puberty, where we can act as mentors (Louv 2005) so that they don't lose their sense of wonder (or decide it is merely 'kid's stuff'). This is the responsibility of the education system, parents and counsellors at schools. One project here of major value is ensuring adolescents can undertake a 'Rite of Passage' in nature (Washington 2018a).
- Lobby the media that it *is* their job to discuss the big issues such as our worldview, our values, our conception of nature, and yes – our sense of wonder.

Conclusion

Rejuvenating our sense of wonder towards nature is important ethically, but also for very practical reasons. Why practical? Well, as an environmental scientist I have written other books (e.g. Washington 2013b, 2015) about the 'practical solutions' that are needed to solve the environmental crisis. However, the majority of these solutions had already been listed by Catton (1982) or in other books by the Ehrlichs (e.g. Ehrlich and Ehrlich 1991) or David Suzuki (e.g. Suzuki and McConnell 1999) or Lester Brown (e.g. 2011). Yet society has *not adopted these solutions*, despite them being based on excellent, rational, environmental science. Instead, society has ignored scientific reality – as well as ecological ethics. I have been searching for the reasons for decades, and have concluded there are two prime causes (Washington 2018a):

1)   Society's addiction to an impossible and unsustainable *endless growth myth*.
2)   The worldview of *anthropocentrism* and human supremacy.

I have spent decades seeking to change both. The first by involvement in the Center for the Advancement of a Steady State Economy (CASSE, see: www.steadystate.org/) (e.g. Washington 2014, 2017; Washington and Twomey 2016). The second via activism for ecocentrism (e.g. Washington et al. 2017, 2018; Kopnina et al. 2018; Washington 2018a). In this regard let me again suggest *you* might like to sign the 'Statement of Commitment to Ecocentrism' (EC 2018).

Clearly, society needs to abandon the worldview of anthropocentrism, as this is a dead end, one that is actually wounding the planet in a huge extinction event (see Chapter 1). Our 200-year experiment of pretending we are separate from (and better than) nature has ended in disaster. Every environmental indicator shows this. It is time to accept that failure – and try something else. We thus as a society desperately need ecocentrism – but how do we transform worldview quickly? My solution is to rejuvenate our sense of wonder, as I believe only through this are we likely to find the *care and deep belief* that allow us to change, to undertake the 'Great Work' of healing the world (Berry 1999). Through rejuvenating our sense of wonder we can heal the rift between ourselves and nature; we can come home.

What can *I* do?

There are various things we can do to bring wonder back into our lives. One of the most important is to rejuvenate our personal sense of wonder, through the following (Washington 2018a):

1) *Be there with nature.* Get out into nature every now and again, into wild places so your spiritual batteries can be recharged and you can re-experience the world in wonder. Wilderness and wild places are important not just because of the wealth of native species they retain (and which they allow to continue evolving); they are essential as they are repositories of the creative impulse, of imagination, of inspiration, of mystery. When I enter wilderness I feel as if my soul expands to touch the infinite and eternal, and I become 'more' than I was before (Washington 2018a).

2) *Take your children to wild places* so they can see the natural world as it really is and bond with it. Go hiking, point out the fascinating aspects of our world (the more you know the more you see). Take up nature photography or drawing, write poetry, go on retreats to wild places, meditate in a beautiful natural spot – all of these let one contemplate the beauty of the natural world and experience harmony with it. By bonding with the land again we connect with a reality that beckons us, which is larger and other than ourselves (Tacey 2000). Quite simply we belong.

3) Take time to *ponder* – whether this is called meditation or prayer or contemplation or 'dadirri', or just sitting somewhere at one with the world. This means healing our short attention span. We should also make sure children have some period in their calendar where they too experience nature, on nature's time (i.e. no computer and no TV). Nature rituals can assist here to rejuvenate wonder (see Chapter 7; Washington 2018a).

4) Keep *your imagination*, creativity and artistic expression alive. In these you find the well-spring of your 'being' which renews your sense of wonder. Take up or resurrect an art or craft or poetry so you can express the creativity inside you. Take part in the artistic life of your community.

5) Cherish the *imagination of your children* and youth in general, especially so they can survive the turmoil of puberty.

6) *Encourage your empathy* on a sunny day. Find a beautiful spot and let your defences down and empathise with the natural world. Perhaps you too will

find, as Thoreau (1854: 86) did, that: 'Every little pine needle expanded and swelled with sympathy and something kindred to me'.

All the above are very relevant to reinventing yourself to heal (see Chapter 7). If we rejuvenate our sense of wonder towards nature then we become better healers.

References

Abram, D. (1996) *The Spell of the Sensuous*, New York: Vintage Books (Random House).
Abram, D. (2010) *Becoming Animal: An Earthly Cosmology*, New York: Vintage Books.
Assadourian, E. (2013) 'Re-engineering cultures to create a sustainable civilization', in *State of the World 2013: Is Sustainability Still Possible?*, ed. L. Starke, Washington: Island Press, pp. 113–125.
Baccelard, G. (1969) *The Poetics of Space*, New York: Beacon Press.
Berry, T. (1988) *The Dream of the Earth*, San Francisco: Sierra Club Books.
Berry, T. (1999) *The Great Work*, New York: Belltower.
Brown, L. (2011) *World on the Edge: How to Prevent Environmental and Economic Collapse*, New York: W.W. Norton and Co.
Carson, R. (1965) *The Sense of Wonder*, New York: Harper-Row.
Catton, W. (1982) *Overshoot: The Ecological Basis of Revolutionary Change*, Chicago: University of Illinois Press.
Clark, R. (1971) *Einstein: The Life and Times*, Cleveland, OH: World Publishing.
Collins, P. (2010) *Judgment Day: The Struggle for Life on Earth*, Sydney: UNSW Press.
Corcoran, P. (2001) 'John Muir', in *Fifty Key Thinkers on the Environment*, ed. J. Palmer, London: Routledge, pp. 131–136.
Crist, E. (2012) 'Abundant Earth and the population question', in *Life on the Brink: Environmentalists Confront Overpopulation*, eds. P. Cafaro and E. Crist, Georgia: University of Georgia Press, pp. 141–151.
Curry, P. (2011) *Ecological Ethics: An Introduction*, 2nd ed. Cambridge: Polity Press.
Daly, H. (1991) *Steady State Economics*, 2nd ed. Washington: Island Press.
Daly, H. (2014) *From Uneconomic Growth to the Steady State Economy*, Cheltenham: Edward Elgar.
Daly, H. and Cobb, J. (1994) *For the Common Good: Redirecting the Economy toward Community, the Environment, and a Sustainable Future*, Boston: Beacon Press.
Dietz, R. and O'Neill, D. (2013) *Enough Is Enough: Building a Sustainable Economy Is a World of Finite Resources*, San Francisco: Berrett-Koehler Publishers.
Dillard, A. (1992) *Teaching a Stone to Talk*, New York: Harper Perennial.
EC. (2018) Statement of Commitment to Ecocentrism, see: www.ecologicalcitizen.net/statement-of-ecocentrism.php?submit=Read%2Fsign+ecocentrism+statement. (accessed 27 March 2019).
EE. (1990) 'Preserving & cherishing the earth: An appeal for joint commitment in science & religion', *Earth Renewal website*, see: http://earthrenewal.org/open_letter_to_the_religious_.htm (accessed 27 March 2019).
Ehrlich, P. (1986) *The Machinery of Nature*, New York: Simon and Shuster.
Ehrlich, P. and Ehrlich, A. (1991) *Healing the Planet: Strategies for Resolving the Environmental Crisis*, New York: Addison-Wesley Publishing Company.

Emerson, R. W. (1836) 'Nature', in *The Portable Emerson*, eds. C. Bode and M. Cowley, (1981), London: Penguin Books, pp. 7–50.

Evernden, N. (1985) *Natural Alien: Humankind and Environment*, Toronto: Toronto University Press.

Evernden, N. (1992) *The Social Creation of Nature*, Baltimore and London: John Hopkins University Press.

Fisher, A. (2013) *Radical Ecopsychology: Psychology in the Service of Life*, Albany: State University of New York Press.

Hempel, M. (2014) 'Ecoliteracy: Knowledge is not enough', in *State of the World 2014: Governing for Sustainability*, ed. L. Mastny, Washington: Island Press, pp. 41–52.

Hill, S. B. (2001) 'Working with processes of change, particularly psychological processes, when implementing sustainable agriculture', in *The Best of … Exploring Sustainable Alternatives: An Introduction to Sustainable Agriculture*, ed. H. Haidn, Saskatoon: Canadian Centre for Sustainable Agriculture, pp. 125–134.

InterculturalStudies (2010) Margaret Mead quote. See: www.interculturalstudies.org/Mead/biography.html (quote used with permission from Sevanne Kassarjian). (accessed 27 March 2019).

Knudtson, P. and Suzuki, D. (1992) *Wisdom of the Elders*, Sydney: Allen and Unwin.

Kopnina, H., Taylor, B., Washington, H. and Piccolo, J. (2018) 'Anthropocentrism: More than just a misunderstood problem', *Journal of Agricultural and Environmental Ethics*, **31** (1): 109–127.

Leopold, A. (1949) *A Sand County Almanac, with Essays on Conservation from Round River*, New York: Random House (1970 printing).

Louv, R. (2005) *Last Child in the Woods: Saving Our Children from Nature-Deficit Disorder*, London: Atlantic Books.

Louv, R. (2011) *The Nature Principle: Reconnecting with Life in a Virtual Age*, Chapel Hill, NC: Algonquin Books.

Merleau-Ponty, M. (1962) *Phenomenology of Perception*, London: Routledge and Kegan Paul.

Mill, J. S. (1857) *Principles of Political Economy*, Cambridge: Parker and Son.

Nabhan, G. and Trimble, S. (1994) *The Geography of Childhood: Why Children Need Wild Places*, Boston: Beacon.

Neidjie, B., Davis, S. and Fox, A. (1985) *Kakadu Man: Bill Neidjie*, Sydney: Mybrood P/L Ind (Allan Fox and Associates).

Oelschlaeger, M. (1991) *The Idea of Wilderness: From Prehistory to the Age of Ecology*, New Haven/London: Yale University Press.

Rolston III, H. (2012) *A New Environmental Ethics: The Next Millennium for Life on Earth*, New York: Routledge.

Snyder, G. (1990) *The Practice of the Wild*, New York: North Point Press.

Suzuki, D. (1998) *Earth Time*, Sydney: Allen and Unwin.

Suzuki, D. and McConnell, A. (1999) *The Sacred Balance: Rediscovering Our Place in Nature*, Sydney: Allen and Unwin.

Tacey, D. (2000) *Re-enchantment: The New Australian Spirituality*, Sydney: Harper-Collins.

Taylor, B. (2010) *Dark Green Religion: Nature Spirituality and the Planetary Future*, Berkeley: University of California Press.

Thomashow, M. (1996) *Ecological Identity: Becoming a Reflective Environmentalist*, Cambridge, MA: MIT Press.

Thoreau, H. D. (1854) *Walden: Or Life in the Woods*, New York: Dover Publications (current edition 1995).

Vetlesen, A. (2015) *The Denial of Nature: Environmental Philosophy in the Era of Global Capitalism*, London: Routledge.

Washington, H. (2002) *A Sense of Wonder*, Rylstone: Ecosolution Consulting.

Washington, H. (2013a) *Poems from the Centre of the World, lulu.com*, see: www.lulu.com/shop/haydn-washington/poems-from-the-centre-of-the-world/paperback/product-21255751.html. (accessed 27 March 2019).

Washington, H. (2013b) *Human Dependence on Nature: How to Help Solve the Environmental Crisis*, London: Earthscan.

Washington, H. (2014) *Addicted to Growth?*, Sydney: CASSE NSW, see: https://steadystatensw.files.wordpress.com/2015/01/addictedtogrowthdocfinalboxesprintfinaloct17th.pdf. (accessed 27 March 2019).

Washington, H. (2015) *Demystifying Sustainability: Towards Real Solutions*, London: Routledge.

Washington, H. (2017) *Positive Steps to a Steady State Economy*, Sydney: CASSENSW, see: https://steadystatensw.files.wordpress.com/2017/06/posstepsroyal11ptjustheaderfinaljune12thebooklowres.pdf. (accessed 27 March 2019).

Washington, H. (2018a) *A Sense of Wonder Towards Nature: Healing the World through Belonging*, London: Routledge.

Washington, H. (2018b) 'Education for wonder', *Education Sciences*, **8** (3): 125. DOI:10.3390/educsci8030125.

Washington, H., Chapron, G., Kopnina, H., Curry, P., Gray, J. and Piccolo, J. (2018) 'Foregrounding ecojustice in conservation', *Biological Conservation*, **228**. 367–374.

Washington, H., Taylor, B., Kopnina, H., Cryer, P. and Piccolo, J. (2017) 'Why ecocentrism is the key pathway to sustainability', *The Ecological Citizen*, **1**: 35–41.

Washington, H. and Twomey, P. (2016) *A Future beyond Growth: Towards a Steady State Economy*, London: Routledge.

Watts, A. (1973) *The Book on the Taboo against Knowing Who You Are*, London: Abacus.

Weber, A. (2016) *The Biology of Wonder: Aliveness, Feeling, and the Metamorphosis of Science*, Gabriola Island, BC, Canada: New Society Publishers.

10

IS COLLAPSE COMING?

Yet upon waking
I practice acceptance.
What's coming is coming,
Whether tragedy or farce
So why rail against fate?
Is not the point
To do the right thing
Even if we fail?
And what greater value
Can there be
Than loving more and more
A wondrous beauty
That is now slipping away?

From 'Upon Waking' (Washington 2018a)

SUMMARY

This chapter asks a question we are often told not to ask – 'Is collapse coming?' Some would say this is 'doom-mongering', will cause panic, will switch readers off, or is defeatist. However, if we are going to abandon denial and accept reality, then it is a question we *should* ask, in the light of the grim predicament discussed in Chapter 1. The chapter first discusses the meanings of 'crisis', 'collapse' and 'catastrophe'. Various statements about the collapse of civilisation are discussed, most suggesting that this is likely, some arguing it is starting right now. However, the chapter points out that 'collapse' can mean many things. Most meanings are for societal collapse, but even then it could mean many things (six are discussed). The terms 'long emergency' and 'long descent' are then discussed, with the latter arguing collapse will come in steps (and this may prove positive, allowing healing at each step). Given most scholars focus on societal collapse, the chapter then covers its less-discussed twin: ecosystem collapse. The chapter then covers 'Is it too late? Optimism, pessimism and *realism*', a key discussion in the light of how many people become apathetic, thinking mistakenly that it is too late. The chapter argues passionately that 'It is never too late!' and that we need to adopt a realism that then acts on the positive (if challenging) solutions that exist. The chapter concludes by discussing 'Towards a good collapse'. This concludes that a major reduction in society's impact on nature is needed, and this could be called a 'collapse'. However, it could also be called a 'transformation' to a sustainable society. It lists seven ways in which a collapse could be considered 'good'. For the detail explaining this summary, please read on; however, if you just want to know the things *you can do* (six are listed) about this, then turn to the end of the chapter.

Introduction

One is continually being told 'don't talk about doom and gloom!' and 'don't talk about crisis or collapse!' However, I think this is a topic that has to be discussed. So … is collapse coming? Well, the short answer is *maybe*. The long answer is that statistically one would have to say *probably*. Why do I say this? Well, as Chapter 1 has shown, the trend shown by environmental indicators can only be described as 'grim'. Things are getting worse rapidly, for many problems are worsening exponentially. As Rees (2017) notes, politicians should try to understand that: 'there is *always* a conflict between human population/economic expansion and "protection of the environment"'. However, that is not to say that we cannot solve the environmental crisis and move to a sustainable future (Washington 2015), because we still *can*, even after three decades of denial and doing very little (Robinson 2013). The big question remains: 'Will we move urgently to undertake the known solutions needed?' It is indeed an excellent question. As we shall see, the answer is still in play, it will depend on you, me and the rest of humanity – and whether we really do act to heal the world. We also need to think through what collapse actually means. As we shall see, there may be such a thing as a 'good' collapse.

Collapse, crisis, catastrophe?

The terms 'crisis' and 'collapse' and 'catastrophe' are often used, but are these valid words to use? The *Oxford English Dictionary* defines *crisis* as:

A time of intense difficulty or danger; a time when a difficult or important decision must be made.

Given the summary in Chapter 1, today is clearly a time of intense difficulty in terms of the natural world and the gifts it gives society. It is also a time when a difficult and important decision (or series of decisions) must be made to maintain the ecosystems that maintain our society. In 1982 it was argued that humanity is in a 'predicament' not a crisis, since the conditions we face are not of recent origin, and will not soon abate (Catton 1982). However, since then things have got worse, and I believe the term environmental 'crisis' is also a correct word to use when half the world's species are at risk of extinction and ecosystems are being degraded and collapsing (see later). Humanity is thus in a predicament where we face a major environmental crisis.

The *Oxford Dictionary* defines 'catastrophe' as: 'an event causing great and usually sudden damage or suffering; a disaster'. It then gives an example, being 'an environmental catastrophe'. Given the extent of mass extinction, soil erosion, industrial pollution, growing population and consumption, growing and unsustainable water use, again one can validly call this a catastrophe, or at least a 'catastrophe-in-the-making' (Washington 2015).

But what *is* a 'collapse', and have other societies suffered such? Most people mean by collapse a *societal* collapse, which can also be called the fall of a complex human society or civilisation. Generally, collapse of a civilisation (societal collapse) is considered to involve a major reduction in population (Diamond 2005). Sometimes this collapse may be by revolution and war, other times by famine and plague, or other environmental disasters caused by degrading the ecosystems on which we depend (Ibid.). Biologist Jared Diamond (2005) in his book *Collapse* shows that societies that deny or ignore their problems do indeed collapse and lists historical examples. He also notes that our modern society has more drivers pushing us towards collapse than ancient societies had (not a good situation to be in). Denial of some problems can thus be not only life-threatening but even society-threatening. As Catton (1982: 213) noted: 'But believing crash can't happen to us is one reason why it will'.

But is collapse likely? Kunstler (2005) argued a comprehensive downscaling, downsizing and relocalising of activities in society *must happen*, whether we do so voluntarily or are dragged kicking and screaming into such a process. He noted the big question is how much disorder must we endure as things change? Greer (2009) concludes that civilisation is sleepwalking towards the abyss, as we fail to grasp the most basic elements of ecological reality. Eminent environmental scientist Paul Ehrlich (2013) has said he gives humanity a '10% chance' of solving our predicament, indeed Ehrlich recently concluded that collapse of civilisation was a 'near certainty' within a few decades (Carrington 2018). Scientist Graham Turner (2014a), who modelled world data for the 'Limits to Growth' (LTG) model (Turner 2008, 2014b) commented to me that collapse is coming, and we have to face this and work out what to do. Turner (2014b: 16) states:

'Regrettably, the alignment of data trends with the *LTG* dynamics indicates that the early stages of collapse could occur within a decade, or might even be underway'. Rees (2019: 144) notes that: 'the era of material growth will soon end either through systems implosion or a planned descent'. He notes (p. 145): 'it is time for the world to admit that continued adherence to prevailing socially-constructed illusions risks fatal catastrophe'. Martenson (2019) argues:

> *Collapse is already here* ... However, unlike Hollywood's vision, the early stages of collapse cause people to cling even tighter to the status quo. Instead of panic in the streets, we simply see more of the same – as those in power do all they can to remain so, while the majority of the public attempts to ignore the growing problems for as long as it possibly can.

Many climate and environmental scientists are worried that we are coming close to overheating the world due to climate change (Hansen et al. 2013; Spratt and Dunlop 2018; Rees 2019). UN Secretary-General Antonio Guterres recently warned that the world is facing 'a direct existential threat' and must rapidly shift from dependence on fossil fuels by 2020 to prevent 'runaway climate change' (Lederer, 2018). We could in fact go beyond the tipping point into a runaway phase that heats the world to the point of social and ecological collapse (Sherwood and Huber 2010; Hansen et al. 2013). However, the key question here is: 'What do we mean by collapse?' Collapse *could* be:

1. Financial collapse when the huge existing debt bubble of $247 trillion (Bloomberg 2018) bursts and many banks and financial institutions collapse (likelihood = strong).
2. A collapse in energy and material use, possibly caused by a financial collapse (probability = medium).
3. Collapse of many ecosystems, including coral reefs, some rainforests, wetlands and mangroves and the extinction of a large part of the life on Earth, possibly up to (or exceeding) half of life (likelihood = quite strong sadly if we fail to act quickly).
4. A climate collapse where the world exceeds a 2 degree temperature rise, which may lead to runaway climate change and a possible 5 degree rise in temperature (Hansen et al. 2013). This is likely to lead to the collapse of food systems and native ecosystems, and to cause massive flooding, extreme weather, extreme heat waves, major disease outbreaks and massive water shortages (see Figure 1.2; Pittock 2009; IPCC 2014; UCS n. d.) (likelihood = medium, note that we may not know we have actually crossed a 'tipping point' until well *after* we have crossed it).
5. Societal and ecological collapse caused by a limited (or more extensive) nuclear war, possibly exacerbated by a food and water crisis caused by climate change. This may cause a 'nuclear winter' that causes widespread famine (Robock and Toon 2010) (likelihood = low to medium).

6. A collapse as shown by Hollywood blockbuster movies such as *2012* or *The Day After Tomorrow* (likelihood = very remote).

Collapse could thus come in many forms (I am sure you can think of other scenarios). The probabilities above are my personal estimates and may well be wrong (and of course their probability will depend on what healing actions society takes to avert these). One thing we can be certain of is that the future will not be quite 'what we expect'. Collapse may come with a crunch, the 'perfect storm' caused by collision of changing climate; spreading ecological disorder (including soil loss, deforestation, water shortages, species loss, ocean acidification); population growth; unfair distribution of the costs, risks and benefits of economic growth; ethnic and religious tensions; and proliferation of nuclear weapons (Orr 2013: 279). Alternatively, it may be an ongoing decline that continues to relentlessly degrade our world, so that each new generation suffers from 'baseline shift' and fails to realise just how much has already been lost or degraded (Pauly 1995).

So collapse may well happen, and I shall discuss later that *in some form* this is actually needed. It depends on the form. The key question should not be whether some form of collapse happens but: 'How much damage will we let happen?' In other words, to what extent do we allow collapse to turn into catastrophe? We should also ask: 'How much of the Earth will we save?' How much of the biosphere (that supports us all) will be saved – that is the real question (Robinson 2013). Can we prevail over our destructive private interests, when the common good of humanity and the biosphere is at stake? It remains an open question (Ibid.). In the end it will be up to *us*. There is nobody else. The key thing to consider here is that the more we actively work to transform our society towards true sustainability, the less serious will be the collapse (and we may thus avoid catastrophe).

The long emergency and long descent?

Whether we solve the environmental crisis or not will be played out this century, almost certainly before mid-century (Meadows et al. 2004), though healing solutions would of course have to extend beyond that. Healing our predicament won't be over in a moment, and in fact it is likely to be the 'long emergency' (Kunstler 2005; Orr 2013). We will need to act for many decades to heal the world, perhaps centuries. Our possible future has also been called the 'long descent' (Greer 2008), a form of decline or collapse (as our energy and material use *must* decrease). Greer argues that past civilisations that collapsed did so in stages, so ours probably will too. I would suggest that hopefully we can learn about sustainability and healing at each step in the descent? This long descent collapse could actually also turn out to be the *long ascent of healing* where we sustainably transform our society. The key thing to understand is that an 'all at once' collapse of everything is quite unlikely, it is far more likely that things will continue to get increasingly worse, and parts of our unsustainable civilisation will fail at various times. A collapse of the financial system (which is quite likely) might wake up society to undertake the solutions needed. However, we should bear in mind that the Global Financial Crisis (GFC) *failed* to wake up society, indeed served only to entrench the idea that we had to 'grow our way' past the GFC (Clements 2016). So if collapse is coming, it is likely to be a long collapse and go on for decades. This makes it even more important that each of us (and our descendants) seeks to heal the planet.

Ecosystem collapse

Most uses of the term 'collapse' come from a meaning of societal collapse; however, ecosystem collapse is just as important (Washington 2013). Ultimately, human existence depends on maintaining the rich web of life within which we evolved. The rapidity of current environmental change is *unique in human history* (Gowdy et al. 2010). Many entire ecosystems are degraded and verging on collapse (MEA 2005; Kumar 2010). If you keep removing the rivets that hold together an aeroplane, at some point it will fall apart and crash (Ehrlich and Ehrlich 1981). If you keep removing species, so will ecosystems. Historically, there are no recorded cases where sudden unexpected losses of biodiversity in major ecosystems have been to the benefit of humanity (Pascual et al. 2010). We know what happens to a house if you take out too many of its foundations: it collapses. The same thing can happen with ecosystems, and the term is fully 'appropriate' to our situation, so we shouldn't hide this under jargon (Washington 2015). However, scientists often *do* use jargon, and one sees descriptions of 'irreversible non-linear changes' or 'regime shifts' (Kumar 2010). What they really mean is ecosystem collapse. Such collapses can produce large unexpected changes in ecosystem services (i.e. the benefits nature provides society). Examples include massive changes in lakes, degradation of rangelands, shifts in fish stocks, breakdown of coral reefs and extinctions due to persistent drought (MEA 2005). Martenson (2019) correctly notes: 'While it's entirely possible for any one ecological mishap to be due to a natural cycle, it's weak thinking to assign the same cause to dozens of troubling findings happening all over the globe'.

But is ecosystem collapse happening? Chapter 1 summarises many causes of such collapse. The MEA (2005) identified that 60% of ecosystem services are being degraded and found that extinction rates were 1,000 times higher than those in the fossil record (more recent estimates put this at 10,000 times (Chivian and Bernstein 2008). Humanity's ecological footprint has doubled since the late 1970s (Sukhdev 2010) and is now at least 1.7 Earths (GFN 2019). This is the amount of land it would take to sustainably provide for the demands humanity places upon the Earth. Of course we only have *one* Earth, hence why ecosystems are degrading. We have massively overshot what the Earth can sustainably provide (Catton 1982; Wijkman and Rockstrom 2012).

So what causes an ecosystem to collapse? There are many stresses – loss of habitat, introduced species, nutrient pollution, toxification of ecosystems. The climate crisis on top of all the other environmental problems will hugely accelerate ecosystem decline and collapse. An ecosystem may change gradually until it reaches a threshold – and then collapses. There can be 'irreversible regime shifts', and it is impossible to know exactly where the dangerous thresholds lie (Elmqvist et al. 2010). Once an ecosystem has collapsed, recovery to the original state may take decades or centuries, and may sometimes be impossible. Given the current human impacts on the biosphere, the notions of 'restoration and restorability' need to: 'be expressed with many caveats' (Ibid.: 88). In other words, we would be fools to think we can restore all ecosystems that collapse. Some of these non-linear changes can be very large and have substantial impacts on human well-being. The MEA (2005) notes that for most ecosystems, science cannot predict the thresholds where change will occur. Thus we cannot say which particular ecosystems will collapse, at what level of stress, or when (Washington 2015).

Ecosystem collapse has been caused by: eutrophication, overfishing, species introduction, nutrient impact on coral ecosystems, regional climate change, the unsustainable bush-meat trade and loss of keystone species (MEA 2005). Does ecosystem collapse matter to us? There is a real (although unknown) possibility that net biodiversity loss will have catastrophic effects on human welfare (Gowdy et al. 2010). Certainly, we do know that humans are totally dependent on the ecosystem services that nature provides (MEA 2005; Washington 2013). If enough ecosystems collapse, this must impact big time on human well-being. So humanity is dependent on nature, but we have ignored and denied this reality, to our (and nature's) great cost. Of course, as the old bumper sticker observed (given this dependency) – 'Nature bats last'. Ignoring or denying ecosystem collapse is not an option if we are to heal the planet.

Is it too late? Optimism, pessimism and *realism*

If we discuss the likelihood of collapse, we have to consider the question of whether 'It is too late' and what this might mean. We need also to discuss optimism, pessimism and realism, as detailed by Washington (2015). As Chapter 3 discusses, we have to *accept reality* to solve our problems. Daily (1999) argues that ecologist Peter Vitousek got it right when he said:

> We are the first generation with tools to understand the changes in the Earth's systems caused by human activity and the last with the opportunity to influence the course of many of the changes now rapidly under way.

It is time to use those tools to understand our problems, then fix them. It is time to 'grow up' and drop our delusions of being 'Masters of Nature' (Washington 2015). We never 'conquered' nature, we just raided fossil fuels to cause the environmental crisis (Greer 2009). We stole fossil energy from the past, and (due to the consequent impacts) we are now robbing the future, making it unsustainable for those who follow us (Washington 2015). Sustainability demands we reassess ourselves (see Chapter 7). Environmentalism has now become the most significant human, moral and theological problem confronting the contemporary world (Collins 2010). If we don't face up to the ecological crisis, we will have no future as sane, ethical and spiritual beings (Swimme and Berry 1992). It is time to stop being a 'seriously dumb species' (Soskolne 2008: xvii) if we are going to succeed in healing the world.

Humans are intelligent (often) and when committed can (and have) solved huge problems. World War II, the Marshall Plan to rebuild Europe after that war and the Space Race are examples of successful major actions that some would have said were impossible (Washington 2015). We also acted on the problems of acid rain and the hole in the ozone layer caused by CFCs (Ibid.). Atlee (2003) describes 'co-intelligence' as the human capacity to generate creative responses, and it is this we need to foster. We face a huge problem, but it is also a major 'opportunity' to build a truly sustainable future. We should be embracing these solutions for many reasons. For example, we need to stabilise the population for reasons of food security, biodiversity protection and water supply, but it also reduces carbon emissions and improves quality of life, and reduces impacts on the poor (Washington 2015). By acting responsibly on climate change, we can solve the energy crisis and also take a big step towards slowing the extinction crisis. We should stop burning fossil fuels for health reasons as well as stopping climate change (Haswell and Washington 2014).

Robinson (2013: 374–5) concludes:

> [i]f we were to do everything right, starting this year and continuing for the next several decades … decarbonise, and to conserve, restore, protect, replace and so on – then we could do it … it is physically possible … are we going to do everything right in the rest of the twenty-first century? … it looks very unlikely. We just aren't that good, either as a species or a civilization … We are going to do damage in the twenty-first century, possibly big damage.

Neither 'fear of pessimism' (doom and gloom), nor 'dogged determination' to remain optimistic are reasons to understate our predicament and the extent of the crisis we face. I have sat through many talks where the speaker insists how 'optimistic' he or she is, and books likewise (e.g. Edwards 2010). However, Ehrenfeld (1978) notes that optimism is necessary for those who attempt the impossible. In fact humanity has a strong 'optimism bias' where we consistently positively overestimate what the future may bring (Sharot 2011). Optimism and pessimism are equal distractions from what we need: *realism*, a commitment to nature and to each other, and a determination not to waste more time

(Engelman 2013). 'Feeling you have to maintain hope can wear you out' notes Joanna Macy (2012), who advises:

> Just be present, when you're worrying about whether you're hopeful or hopeless or pessimistic or optimistic, who cares? The main thing is that you are showing up, that you're here, that you're finding ever more capacity to love this world, because it will not be healed without that.

Moore and Nelson (2013: 229) beautifully express what we need now – 'moral integrity':

> But to think that hope and despair are the only two options is a false dichotomy. Between them is a vast fertile middle ground, which is integrity; a matching between what we believe and what we do. To act justly because we believe in justice. To act lovingly toward children because we love them. To refuse to allow corporations to make us into instruments of destruction because we believe it is wrong to wreck the world. This is moral integrity. This is a fundamental moral obligation – to act in ways that are consistent with our beliefs about what is right. And this is a fundamental moral challenge – to make our lives into works of art that embody our deepest values.

I would add too that we should make our lives a work of healing. I should emphasise that I am not arguing there is no hope at all, just questioning blind hope. If we accept the reality of our problems and put in place the ten key solutions of the next chapter, there is indeed hope. *People need a small amount of hope*; they need a dream to work towards.

So is it too late? *No, it's never too late!* Any action we take for healing is better than none. Foresters say the best time to plant a tree is decades ago, the second best time is today (Zencey 2013). Just as water can erode a rock, the continuing pursuit of cultural change can add up to much more than the sum of its parts. Seeds sown today may take root when humanity reaches for solutions to rebuild, as systems unravel under the unbearable burden of sustaining a global consumer economy (Assadourian 2013). Ecologist Paul Ehrlich notes that there are thresholds in human behaviour when cultural evolution moves rapidly, so that: 'When the time is ripe, society can be transformed virtually overnight' (in Daily and Ellison 2002: 233). Today could be such a time.

So, we should all accept the invitation to abandon pessimism and depression, which really means one will do nothing to make things better. However, we should equally abandon gung-ho optimism (and denial), which downplays the severity of the mess we have created and pretends that the solutions are simple. They won't be simple, they will be hard, but they *are* possible (Washington 2015). Instead, become a 'realist', accept we have major problems that we must solve to become sustainable. Not in a decade, not in a year. Now. It is never too late to do the right thing and to act for healing. Many who ask 'Is it too late?' are actually looking for an excuse to be apathetic, as they cannot be bothered to put energy into healing the planet (and make endless excuses as to why they do nothing). They need to reinvent themselves (see Chapter 7) to get past this apathy. Even young people fall into this apathy trap. It is not all over and never will be until the last of us ceases trying to heal the world. For everyone, especially young people, there is a living world – and a sustainable future for both humans and nonhumans – to fight for.

The environmental crisis is solvable; we *can* heal our world. Not by pretending and deluding ourselves. Not by saying it's 'too hard'. Not by prostrating ourselves before a neoliberal 'Free Market' god and hoping it will save us. Rather, it will be solved by accepting the need and rising to the challenge: seeking to heal our world (Washington 2015). Meadows et al. (2004) invite us to imagine how different things would be if we accepted a global transition to a sustainable society was possible. It has been argued that making broader change involves three things:

1. A big idea of how things could be better;
2. A commitment to move beyond individual actions and join with others to make it real;
3. Then *real* action must follow (Leonard 2013: 250).

The big idea is, I believe, obvious – *seeking to heal our world*. Working with others should be obvious; however, we need to rejuvenate social capital and the art of dialogue to do this. Taking real actions moves us past denial and apathy into real change. The missing ingredient is collective engagement for political and structural change to heal our world. However, if we move past denial and accept our problems, then we can commence the 'Great Work' (Berry 1999) of healing (see Chapter 11). A meaningful sustainability is thus really healing the planet, along with our society and our maladjusted economy. This may take time. We should accept that true sustainability may not arrive for decades or even centuries (Engelman 2013). However, the next few decades will be crucial. The future we reach together in terms of healing our world will essentially be set in motion *now*.

BOX 10.1 THINK GLOBALLY, ACT LOCALLY: THE MANTRA FOR A SUSTAINABLE PATH, COLIN L. SOSKOLNE

The title of this box provides one answer to the question raised by this book. As a mantra, 'Think globally, act locally' urges people to consider the health of the entire planet and to take action in their own communities and cities to mitigate harmful global changes that are rendering life-supporting ecosystems unsustainable. Long before governments began enforcing environmental laws, individuals were coming together to protect habitats and the organisms that lived within them. These efforts are referred to as grass-roots efforts. They occur on a local level and are primarily run by volunteers and helpers. While this mantra originally began at the grassroots level, it is now a global concept of great importance. It is not only volunteers who take the environment into consideration; it is corporations, government offi-cials, the education system, and local communities.

Unless we embrace these approaches and transition into more local futures, ever-widening disparities will continue to create increasing social dis-ease, leading to a greater burden of preventable morbidity and prema-ture mortality. Current trajectories under the *status quo* of business-as-usual favor the rich becoming richer and the poor relatively poorer, with both cat-egories growing in number. Too small a number of middle-class in a community is, if history bears repeating, a formula destined to lead to a major conflagration. Would it not be wise to work toward reducing this widening gulf and thus prevent a conflagration? Preventing harm is, after all, the mission of public health.

Radical action against the *status quo* is identified by Chris Hedges as our obligation in his 2015 book *Wages of Rebellion: The Moral Imperative of Revolt*. Hedges' message is clear: popular uprisings in the United States and around the world are inevitable in the face of environmental destruction and wealth polarization. Under his view of 'environmental destruction', we include global trends to ever worsening eco-toxicity, population growth, and, in particular, to climate change and global heating.

Unless we begin to value life in all of its forms as embodied in *The Earth Charter* – an ethical framework anchored in global values for building a just, sustainable, and peaceful global society in the 21st century, our destiny is one of grave disruptions to life as we know it and, likely, to more extensive extinctions, including that of the human species. *The Earth Charter* seeks to inspire in all people a new sense of global interdependence and shared responsibility for the well-being of the whole human family, the greater community of life, and future generations. It is a vision of hope and a call to action. It is a 4-page document that provides, in some 59 languages, a clear path forward. It provides the foundational values and ethical framework to help humanity transition to a sustainable path, individually and collectively.

For further information:

www.globalresearch.ca/the-localization-movement-creating-a-viable-local-economy-challenging-the-new-world-order/5371155

www.localfutures.org/?link_id=36&can_id=f84d543de871068b92097cb2b3b f68eb&source=email-prevent-climate-chaos-stop-insane-trade-now&email _referrer=email_515283&email_subject=prevent-climate-chaos-stop-insane-trade-now

http://earthcharter.org/virtual-library2/the-earth-charter-text/

Conclusion: towards a 'good' collapse?

So is collapse coming? Well in some form it almost certainly *is*. In fact a type of collapse is *needed*, as society must step back from its high energy and high material use and massive degradation of nature (Greer 2009). Will financial markets collapse? Almost certainly, as they rely on an endless growth myth which cannot work on a finite planet. Will such a collapse lead to the death of hundreds of millions (or even a billion) people? I certainly hope not, but that will depend on what we do now as a society to heal the planet and transform our society to be truly sustainable. It is important to understand that the world is in crisis, and some form of collapse is coming – but that it *doesn't have to be a catastrophe*. Catastrophe will only become the outcome if we fail to act to heal the planet, if we stay in denial and pretend we can continue endless growth, a fundamentally unsustainable path.

Ecomodernists are fond of saying they seek a 'good Anthropocene' (EM 2015). However, more importantly we should understand that there can be such a thing as a *good collapse*. Such would involve a major reduction in energy and material use, a cessation of clearing native vegetation, entwined with a major change in worldview and ethics. Will we have less things (less junk)? Almost certainly. Is that a bad thing? No it is not. We can live a good and sufficient life; we don't have to live like princes and princesses (see Chapter 5 re Cuba). We don't need to always have 'more and more', that is an unsustainable ideology we have been brainwashed into believing (see Chapters 2 and 5). If Greer (2009) is right in arguing that collapses proceed in stages, then healing the world will be central to every stage. As we face financial collapses, or perhaps the collapse of broad-scale mining, we can change worldview and learn to live sustainably (or rather, we can learn it *again*, as many of our ancestors did and many Indigenous cultures still aspire to do so, Curry 2011).

Healing the planet thus requires we abandon a denial that some sort of collapse is coming. In fact, healing *requires* a degree of collapse of an unsustainable society taking too much from the world. Healing requires we do our utmost to seek a 'good collapse' by actively transforming society. How can a collapse be good? In the following ways:

1) Well, firstly if it can avoid a massive extinction of *both* human and nonhuman life, then it can only be seen as 'good'.
2) Secondly, if it means a major reduction in energy and material use, then that is good, as these drive overdevelopment, pollution and (currently) climate change.
3) Thirdly, if it means a stabilisation then reduction of the population by humane, *non-coercive* means (rather than famine and war), then it has to be good for both humanity and nature (see Chapter 4).
4) Fourthly, if it means a big reduction in overconsumption and hyperconsumerism, with a move back to *thriftiness*, then it has to be good (see Chapter 5).

5) Fifthly, if it means a move to halt the clearing of forests, wetlands and other native vegetation (and instead reforesting and regenerating native communities), it has to be good.
6) Sixthly, if it means a focus on ecologically sustainable agriculture rather than unsustainable agribusiness, it has to be good.
7) And finally, if it means a change of worldview and ethics from anthropocentrism to ecocentrism, and a celebration of wonder towards life (Washington 2018b) – then it has to be especially good.

The above are all possible as well as necessary. However, to achieve 1) is becoming increasingly hard due to society's denial of ecological reality; 2) is possible in part via the rapid expansion of renewable energy, though consumerism holds back material use reduction; 3) is hard because so many institutions in society oppose population control and consider it a taboo; similarly, 4) is hard as corporations and the advertising industry revel in hyper-consumerism (purely due to profit motives); 5) is possible but probably only if society en masse supports the 'Nature Needs Half' or 'Half Earth' visions (Locke 2013; Wilson 2016; Dinerstein et al. 2017); 6) is possible only if society finally wakes up and understands that its food production is currently not ecologically sustainable, and that this requires major and urgent change; 7) of course is possible and would almost certainly be the single most positive thing society could do to create a 'good' collapse. So the question should not be 'Is collapse coming' but rather: 'What can I do to aid a good collapse that can help heal the world?' Indeed, another term for a 'good collapse' would be a *transition to a sustainable future*. We thus need to transform society to become truly ecologically, socially and economically sustainable.

What can *I* do?

There are various things you can do about collapse:

1) Don't be scared by terms such as 'crisis' and 'collapse'. These can be the catalysts that wake society up to make the needed changes so we *avoid catastrophe* for us and for nature.
2) Accept the reality that humanity's use of energy and materials *must decline*, as must our numbers. We can do this deliberately and humanely by supporting a 'good collapse' – a transition to a sustainable future where we actively transform society.
3) Understand that healing the world will be the 'long emergency' that both you and your children will need to work on positively.
4) Point out that it will never be 'too late', and each generation will need to advance the vision of healing the world. This is the 'Great Work' (Berry 1999) that we must all undertake (see Chapter 11).

5) Act as a *realist* who is neither a pessimist nor an optimist. Accept the reality of our predicament – we have major problems we need to solve. Then talk about the positive (if challenging) solutions we have to help heal the world (see Chapter 11).

6) Seek to learn about skills that will be useful in a post-consumer era (where society is in transition). This is not for survivalism in isolation, but to work together for healing. One can learn about wilderness first aid, foraging, blacksmithing, carpentry, cooking, soap-making, candle-making, gardening, midwifery, etc. Such skills are always useful, and one can easily find courses to join. The Resilience website has ideas (www.resilience.org/act-resources/).

References

Assadourian, E. (2013) 'Re-engineering cultures to create a sustainable civilization', in *State of the World 2013: Is Sustainability Still Possible?* ed. L. Starke, Washington: Island Press, pp. 113–125.

Atlee, T. (2003) *The Tao of Democracy: Using Co-Intelligence to Create a World that Works for All*, Manitoba, Canada: The Writers Collective.

Berry, T. (1999) *The Great Work: Our Way into the Future*, New York: Bell Tower.

Bloomberg. (2018) 'Global Debt Topped $247 trillion in the first quarter, IIF says', *Bloomberg*, 11 July 2018, see: www.bloomberg.com/news/articles/2018-07-10/global-debt-topped-247-trillion-in-the-first-quarter-iif-says. (accessed 2 March 2019).

Carrington, D. (2018) 'Paul Ehrlich: 'Collapse of civilisation is a near certainty within decades'', *The Guardian*, 22 March, see: www.theguardian.com/cities/2018/mar/22/collapse-civilisation-near-certain-decades-population-bomb-paul-ehrlich. (accessed 28 February 2019).

Catton, W. (1982) *Overshoot: The Ecological Basis of Revolutionary Change*, Chicago: University of Illinois Press.

Chivian, E. and Bernstein, A. (2008, eds.) *Sustaining Life: How Human Health Depends on Biodiversity. Center for Health and the Global Environment*, New York: Oxford University Press.

Clements, L. (2016) 'Biggest market crash in history is coming as huge debt bubble bursts, Top investor warns', *Sunday Express*, 20 September, see: www.express.co.uk/finance/city/712178/World-set-for-crash-WORSE-than-2008-when-monster-debt-bubble-explodes-warns-investment-ex. (accessed 27 March 2019).

Collins, P. (2010) *Judgment Day: The Struggle for Life on Earth*, Sydney: UNSW Press.

Curry, P. (2011) *Ecological Ethics: An Introduction*, 2nd ed. Cambridge: Polity Press.

Daily, G. (1999) 'Developing a scientific basis for managing earth's life support systems', *Conservation Ecology*, **3** (2): 14.

Daily, G. and Ellison, K. (2002) *The New Economy of Nature: The Quest to Make Conservation Profitable*, Washington: Island Press.

Diamond, J. (2005) *Collapse: Why Societies Choose to Fail or Succeed*, New York: Viking Press.

Dinerstein, E., Olson, D., Joshi, A., et al. (2017) 'An ecoregion-based approach to protecting half the terrestrial realm', *BioScience*, **67**: 534–545. DOI: 10.1093/biosci/bix014.

Edwards, A. (2010) *Thriving beyond Sustainability: Pathways to a Resilient Society*, Canada: New Society Publishers.

Ehrenfeld, D. (1978) *The Arrogance of Humanism*, New York: Oxford University Press.

Ehrlich, P. (2013) *Personal Communication from Prof. Paul Ehrlich at the 2013 Fenner Conference on the Environment 'Population, Resources and Climate Change'*, Canberra, Australia. see: http://population.org.au/fenner-conference-2013-summing.

Ehrlich, P. and Ehrlich, A.. (1981) *Extinction: The Causes and Consequences of the Disappearance of Species*, New York: Random House.

Elmqvist, T., Maltby, E., Barker, T., Mortimer, M., and Perrings, C. (2010) 'Biodiversity, ecosystems and ecosystem services', in *The Economics of Ecosystems and Biodiversity: Ecological and Economic Foundations*, ed. P. Kumar, London: Earthscan, pp. 41–112.

EM. (2015) 'The ecomodernist manifesto', see: www.ecomodernism.org/manifesto-english/ (accessed 2 March 2019).

Engelman, R. (2013) 'Beyond sustainababble', in *State of the World 2013: Is Sustainability Still Possible?* ed. L. Starke, Washington: Island Press, pp. 1–16.

GFN. (2019) 'Country trends', *Global Footprint Network*, see: http://data.footprintnetwork.org/#/countryTrends?cn=5001&type=BCpc,EFCpc (accessed 2 March 2019).

Gowdy, J., Howarth, R. and Tisdell, C.. (2010) 'Discounting, ethics and options for maintaining biodiversity and ecosystem integrity', in *The Economics of Ecosystems and Biodiversity: Ecological and Economic Foundations*, ed. P. Kumar, London: Earthscan, pp. 57–284.

Greer, J. M. (2008) *The Long Descent: A User's Guide to the End of the Industrial Age*, Canada: New Society Publishers.

Greer, J. M. (2009) *The Ecotechnic Future: Envisioning a Post-peak World*, Canada: New Society Publishers.

Hansen, J., Sato, M., Russell, G. and Kharecha, P.. (2013) 'Climate sensitivity, sea level and atmospheric carbon dioxide', *Philosophical Transactions of the Royal Society A*, **371** (2001). DOI: 10.1098/rsta.2012.0294.

Haswell, M. and Washington, H. (2014) 'Not so cheap: Australia needs to acknowledge the real cost of coal', *The Conversation*, 9 June 2014, see: https://theconversation.com/not-socheap-australia-needs-to-acknowledge-the-real-cost-of-coal-26640 (accessed 22 March 2019).

IPCC. (2014) 'Summary for policymakers', *Fifth Assessment Report, International Panel on Climate Change*, see: http://ipcc-wg2.gov/AR5/images/uploads/WG2AR5_SPM_FINAL.pdf (accessed 27 March 2019).

Kumar, P. (2010) *The Economics of Ecosystems and Biodiversity: Ecological and Economic Foundations*, London: Earthscan.

Kunstler, H. (2005) *The Long Emergency: Surviving the End of Oil, Climate Change, and Other Converging Catastrophes of the Twenty-First Century*, New York: Grove Press.

Lederer, E. (2018) 'UN Chief: World must prevent runaway climate change by 2020', *Associated Press News*, see: www.apnews.com/71ab1abf44c14605bf2dda29d6b5ebcc (accessed 2 March 2019).

Leonard, A. (2013) 'Moving from individual change to societal change', in *State of the World 2013: Is Sustainability Still Possible?* ed. L. Starke, Washington: Island Press, pp. 244–252.

Locke, H. (2013) 'Nature needs (at least) half: A necessary and hopeful new agenda for protected areas', in *Protecting the Wild: Parks and Wilderness, the Foundation for Conservation*, eds. G. Wuerthner, E. Crist and T. Butler, 2015, Washington, DC: Island Press, pp. 3–15.

Macy, J. (2012) 'A wild love for the world', *Joanna Macy interview by Krista Tippett, 'On Being', American Public Media*, 1 November 2012, see: https://onbeing.org/programs/joanna-macy-a-wild-love-for-the-world/ (accessed 14 February 2019).

Martenson, C. (2019) 'Collapse is already here', *Resilience*, website 31 January 2019, see: www.resilience.org/stories/2019-01-31/collapse-is-already-here/ (accessed 2 March 2019).

MEA. (2005) *Living beyond Our Means: Natural Assets and Human Wellbeing, Statement from the Board, Millennium Ecosystem Assessment*, United Nations Environment Programme (UNEP), see: www.millenniumassessment.org/documents/document.429.aspx.pdf (accessed 19 Febuary 2018).

Meadows, D., Randers, J. and Meadows, D. (2004) *Limits to Growth: The 30-year Update*, Vermont: Chelsea Green.

Moore, K. D. and Nelson, M.P. (2013) 'Moving toward a global moral consensus on environmental action', in *State of the World 2013: Is Sustainability Still Possible?* ed. L. Starke, Washington: Island Press, pp. 225–233.

Orr, D. (2013) 'Governance in the long emergency', in *State of the World 2013: Is Sustainability Still Possible?* ed. L. Starke, Washington: Island Press, pp. 279–291.

Pascual, U., Muradian, R., Brander, L., et al. (2010) 'The economics of valuing ecosystem services and biodiversity', in *The Economics of Ecosystems and Biodiversity: Ecological and Economic Foundations*, ed. P. Kumar, London: Earthscan, pp. 183–256.

Pauly, D. (1995) 'Anecdotes and the shifting baseline syndrome of fisheries', *Trends in Ecology and Evolution*, **10** (10): 430.

Pittock, A. B. (2009) *Climate Change: The Science, Impacts and Solutions*, Australia: CSIRO Publishing/Earthscan.

Rees, W. (2017) 'What, me worry? Humans are blind to imminent environmental collapse', *The Tyee*, 16 November, see: https://thetyee.ca/Opinion/2017/11/16/humans-blind-imminent-environmental-collapse/.

Rees, W. (2019) 'End game: The economy as eco-catastrophe and what needs to change', *Real World Economic Review*, **87**: 132–148.

Robinson, K. S. (2013) 'Is it too late?' in *State of the World 2013: Is Sustainability Still Possible?* ed. L. Starke, Washington: Island Press, pp. 374–380.

Robock, A. and Toon, B.. (2010) 'South Asian threat? Local nuclear war = global suffering', *Scientific American*, **302** (1): 74–81.

Sharot, T. (2011) 'The optimism bias', *Current Biology*, **21** (23): R941–R945.

Sherwood, S. C. and Huber, M.. (2010) 'An adaptability limit to climate change due to heat stress', *PNAS*, **107** (21): 9552–9555.

Soskolne, C. (2008) Preface to *Sustaining Life on Earth: Environmental and Human Health through Global Governance*, ed. C. Soskolne. New York: Lexington Books, pp. xvii–xviii.

Spratt, D. and Dunlop, I.. (2018) *What Lies Beneath: The Scientific Understatement of Climate Risks*, Melbourne, Australia: Breakthrough - National Centre for Climate Restoration. See: https://climateextremes.org.au/wp-content/uploads/2018/08/What-Lies-Beneath-V3-LR-Blank5b15d.pdf.

Sukhdev, P. (2010) Preface to the *Economics of Ecosystems and Biodiversity: Ecological and Economic Foundations*, ed. P. Kumar, London: Earthscan.

Swimme, B. and Berry, T. (1992) *The Universe Story: From the Primordial Flaming Forth to the Ecozoic Era*, San Francisco: Harper Books.

Turner, G. (2008) 'A comparison of the *limits to growth* with 30 years of reality', *Global Environmental Change*, **18** (3): 397–411.

Turner, G. (2014a) 'Personal comment to the author by Dr Graham M. Turner while the author was organizing the 2014 Fenner Conference on the Environment 'addicted to Growth?''

Turner, G. (2014b) *Is Global Collapse Imminent? an Updated Comparison of 'the Limits of Growth' with Historical Data (research Paper 4)*, Parkville, VIC, Australia: Melbourne Sustainable Society Institute. See: https://is.gd/8dtsWW (accessed 28 February 2018) .

UCS. (n.d.) 'Global warming impacts', *Union of Concerned Scientists*, see: www.ucsusa. org/our-work/global-warming/science-and-impacts/global-warming-impacts (accessed 5 March 2019).

Washington, H. (2013) *Human Dependence on Nature: How to Help Solve the Environmental Crisis*, London: Earthscan.

Washington, H. (2015) *Demystifying Sustainability: Towards Real Solutions*, London: Routledge.

Washington, H. (2018a) *Kinship Poems*, Lulu.com, see: www.lulu.com/shop/haydn-washington/kinship-poems/paperback/product-23580118.html.

Washington, H. (2018b) *A Sense of Wonder Towards Nature: Healing the World through Belonging*, London: Routledge.

Wijkman, A. and Rockstrom, J.. (2012) *Bankrupting Nature: Denying Our Planetary Boundaries*, London: Routledge.

Wilson, E. O. (2016) *Half Earth: Our Planet's Fight for Life*, New York: W.W. Norton.

Zencey, E. (2013) 'Energy as master resource', in *State of the World 2013: Is Sustainability Still Possible?*, ed. L. Starke, Washington: Island Press, pp. 73–83.

11

SEEK TO HEAL: THE 'GREAT WORK' OF TEN KEY SOLUTIONS

And yet …
There remains
A mote of light,
Reminding us
That while tragedy
Lurks all around
Waiting …
Amidst the mourning
Shines the morning,
A hidden joy,
First principle of creation:
Love.

From 'Tragedy' (Washington 2018a)

SUMMARY

This chapter summarises the key points of what 'healing the world' involves: diagnosis and action. It discusses one approach of identifying the 'key solutions' we should focus on vs. the 'pluralism' approach (popular in academia) where every solution is deemed correct. I argue that the pluralism approach wastes energy and actually slows healing. The 'Great Work' of healing is explained. Ten solution frameworks are listed, being: 1) worldview and ethics; 2) reinventing ourselves to heal; 3) overpopulation; 4) consumerism and the growth economy; 5) solving climate change; 6) keeping life alive; 7) appropriate technology: a renewable future; 8) reducing poverty and inequality; 9) education and communication; and 10) the politics of healing.

> These are the solutions that underpin the healing of our world – which we can all be part of. For the detail explaining this summary, please read on; however, if you just want to know the things *you can do* (ten are listed) about this, then turn to the end of the chapter.

Introduction

This book is about healing the world. It is based on the premise that the world needs healing, that this can still be done practically and that we *should* do it ethically. The key turning point is for people to set aside past ideas of 'mastery' and human supremacy – and instead seek to heal a wounded world.

How do we heal?

How do we heal? Well, firstly it involves diagnosis – 'seeing' the problems. That means seeing the reality of the predicament we are in (see Chapters 1 and 3). As we have seen in Chapter 3, humanity overall is not good at 'seeing' or accepting the problems that beset it. Essentially, we ignore and deny them, and keep pretending that (on a finite planet) we can keep growing physically forever. Some in society even pretend we face an exciting world of 'ever more!' This is neither rational nor sustainable; in fact it is 'hubris', a form of arrogant madness (see Chapters 2 and 7) that threatens nature, society and any chance of a harmonious future (Washington 2018b). So, diagnosis is essential in any healing medicine, just as it is in healing the world. Denying the problem in both endeavours just leads to decline and death.

Secondly, healing means we need to have an idea of 'what to do' to make things better. After all, in medicine there is a history of quacks and dangerous 'remedies' – as well as remedies that do work. Similarly with healing the world, there are thousands of solutions proposed out there (some of which contradict each other) – and they cannot *all* be correct. Hence it is essential to winnow out the make-believe fantasies, the denial of reality and the wishful thinking. One approach to solutions is called *pluralism* and comes from a postmodern perspective (as discussed by Washington 2015). This argues that there are many different visions, approaches and pathways that we should be acting on for sustainability. All are often portrayed as 'equally valid'. On the face of it, the claim that there are many different paths and visions sounds great. It means that everyone's approach is correct, so you don't have to worry about deciding which ones to focus on. Under pluralism, everyone can congratulate each other that their particular solution is also 'right'.

Others disagree; they think we should be advocating the *best* vision and the most practical solutions to solve the problems. Jickling (2009) argues against the dominant view in post-structuralism within academia that education has no meaning other than those subjectively ascribed to it. Wals (2010: 150) notes there is a conflict between a deep concern and urgency about the state of the planet, and a conviction in some of academia that it is: 'wrong to persuade, influence or even educate people towards pre- and expert-determined ways of thinking and acting'. Rickinson (2003) and Chawla and Cushing (2007) call for clear objectives and ends in order to make education for sustainable development effective in addressing environmental problems. Kopnina (2012: 701) notes that:

> [r]esearch on education for sustainable development is dominated by calls for pluralistic, emancipatory or transactional forms of education that encourage co-creation of knowledge … and encourage multiple perspectives and critical dialogue on the very concept of sustainable development and education for sustainable development (Gough and Scott 2007; Wals 2007).

Kopnina (2012: 712) concludes that the pluralistic approach to education for sustainable development: 'signals scholarly departure from the "real world" dilemmas concerned with environmental degradation'. She observes (Ibid.) that academic relativism about education for sustainable development: 'might in fact be undermining the efforts of educating citizens in the importance of valuing and protecting the environment'. She notes that scholars of sustainable development seem increasingly engaged in academic debates about the: 'dangers of dogmatic thinking'. She wonders if such theorists might be failing to: 'see the (still standing) forest behind the (receding) trees' (Ibid.).

I share that deep concern. Hence I argue here for a framework of ten *specific solutions* we need to focus on. In regard to healing the world, for almost three decades society has fiddled while Rome has burned around us (Washington 2015). It seems also that many in academia have been playing along in the string section. The dominance of postmodernism within academia has led to arguments for pluralistic approaches, and to a belief that any overarching 'grand narrative' was wrong and should be opposed (as discussed by Butler 2002). I believe this is why so much of the debate around 'sustainability' and 'sustainable development' has become tokenism (Washington 2015). This is why Gare (1995) noted that postmodernism was good at analysing the problems of modernism, but powerless to provide solutions. I believe we *do* in fact need a 'grand narrative' of healing the planet now, if we are to solve the environmental crisis. A 'grand narrative' after all is a 'great story', and stories have always been important in human societies (Abram 2010). Nabhan and Trimble 1994: 21) report a Cibecue Apache elder conclude: 'Stories go to work on you like arrows. Stories make you live right'. However, our grand narrative cannot remain one of endless growth (as was that of 'Our Common Future', WCED 1987). It cannot remain one of refusing to accept ecological limits and physical laws. It cannot remain one of ignoring the centrality of worldview and ethics (Washington 2015, 2018b).

I don't accept the pluralist argument that all solutions are 'equally valid', or that it's fine if some of those solutions contradict others. This promotes dithering and is a massive waste of time and energy. Washington (2015) listed nine key solutions that underpin the 'Great Work' of healing the world. I have added one more here ('Keeping life alive' – given it really needs specific emphasis) to create ten solution frameworks of key relevance for healing our planet. There is good agreement around these ten solution frameworks by environmental scientists and scholars over several decades. Many books have been written about our predicament. Some notable ones are Catton (1982) *Overshoot*, Ehrlich and Ehrlich (1991) *Healing the Planet* Berry (1999) *The Great Work*, Soskolne (2008) *Sustaining Life on Earth* and Brown (2011) *World on the Edge*. It is notable how much agreement there is in such books as to both the problems and the solutions. It's not like the solutions have not been pointed out; it is that they have *not been acted on* (Washington 2015).

Edwards (2005) summarises many of the policies and principles that have been developed on sustainability in many countries. These include the Ontario ORTEE principles; the excellent Netherlands National Environmental Policy Plan; the Natural Step; the Houston Principles; the CERES principles; the Forestry Stewardship Council principles; the Asilomar Declaration for Sustainable Agriculture; the Hannover Principles; Todd's principles of Ecological Design; the Sanborn Principles; the LEEDing Edge; Deep Ecology Principles; Permaculture Principles; and of course the inspirational 'Earth Charter'. He points out there are remarkable similarities in the key values expressed in these principles. He concludes there are seven common themes: 1) *stewardship* (an ecological ethic); 2) *respect for limits* (living within nature's means and preventing waste and pollution); 3) *interdependence* (connectivity within nature, on which humanity depends); 4) *economic restructuring* (that safeguards ecosystems, though rarely reported as going as far as a steady state economy); 5) *fair distribution* (equity); 6) *intergenerational perspective* (taking a long-term view); and 7) *nature as a model and teacher* (which includes respecting the 'Rights of Nature') (Ibid.). These themes are indeed crucial to the healing of our world. If we just acted on these themes alone, it would be a big step forward. In fact all of them are part of the ten solution frameworks discussed later. However, there are other things we must also do for healing, as we shall see.

Accordingly, I believe there *is* a coherent vision of healing the world, which is accepting reality, accepting society is now hugely *un*sustainable and undertaking the 'Great Work' (Berry 1999) of moving to a truly sustainable future (ecologically, socially and economically in the long term). Under that vision there are of course numerous goals, strategies and parts to the solutions. My conclusion concerning pluralism will clearly be unpalatable to some in academia. However, I write here for the educated layperson who seeks to encourage the actual *effective* healing of our world. I seek to provide real solutions to real problems, rather than theorise (under the current latest theory, Washington 2018b), procrastinate, waffle or deny. Each reader must decide for themselves whether the solutions listed are too 'radical' or actually a useful way forward. I believe there are ten clear solutions to heal the world and list these frameworks. Most of these have also been discussed in other chapters. However, none of these solutions is likely to be put into practice *unless* society commits to the overall vision of the 'Great Work' of healing our world.

The 'Great Work'

This book has shown that the worldviews of anthropocentrism, modernism and consumerism have failed to bring a better world (see Chapter 2); in fact they are the ideologies that have wounded our world. They have broken our society, as well as our economy, and are well on the way to breaking the ecosystems on which civilisation relies (Washington 2015). However, most of us in Western society still deny this, but if we could overcome this denial, then we could start *fixing things*. Theologian Thomas Berry (1999) argues that contemporary history has shown three things:

- The devastation of the Earth;
- The incompetence of religion and cultural traditions to deal with this devastation;
- The rise of a new ecological vision of the Universe.

We now need an ecocentric ethic or 'Earth ethics' (Rolston 2012), an ethic not based just on enlightened self-interest, but one that considers the rest of nature too (Assadourian 2010; Curry 2011). Berry (1999) argues our task should be to undertake 'The Great Work' and provides an ecocentric vision for this work of Earth repair or healing. He notes (p. 1) that every culture produces a 'Great Work', an overarching vision that: 'gives shape and meaning to life by relating the human venture to the larger destinies of the Universe'. Macy (2012) seeks a similar vision in her 'Great Turning', while Rees (2010) calls this 'Survival 2100'. This is a very necessary philosophical and ethical 'grand narrative'. It is an overarching worldview, a narrative that provides meaning at a time when we desperately need meaning to foster the deep beliefs that will support us in healing the world (Washington 2015). It is a dream that people can believe in, a dream that can inspire both young and old. It is a dream that can lead to a sustainable future.

Martin Luther King inspired people through his statement 'I have a dream!' He would not have inspired them if he had said 'I have a catastrophe!' People need a degree of hope; they need a dream to work towards. Hope is linked to imagination, and this allows us to conceive the world differently and create a sustainable future (Collins 2010). Change comes from necessity, hope, realisable aspirations and joy, not shame and blame (Wackernagel and Rees 1996). Given the urgency involved in our predicament, this grand narrative, this Great Work of healing the Earth, is precisely the vision we need to move forward to an ecologically sustainable future (Washington 2015). The 'Great Work' is our path to healing the world, and we won't get there unless we move past denial – and as a result take that path. There will be many parts to the Great Work (many are listed below), and they will change over time. However, the big shift is accepting this new task of healing: *the Great Work of Earth Repair*.

Ten solution frameworks

These are adapted from the nine solution frameworks in *Demystifying Sustainability* (Washington 2015). Many of the framework themes have already been discussed in detail in other chapters, so this section serves as summaries of why these ten frameworks are central for healing our world.

1) Worldview, ethics, values and ideologies

Chapter 2 showed the Western anthropocentric and modernist worldview to be broken; it can even validly be called 'evil', as it is anti-life and fundamentally unsustainable (Collins 2010). We desperately need to change our worldview to ecocentrism and our ethics to ecological ethics (Curry 2011; Rolston 2012; Washington et al. 2017; Washington 2018c). Humanity is faced with a complex problem of cultural evolution to create a thoughtful, ecologically literate human, motivated by a: 'sense of responsibility to the planet' (Soskolne et al. 2008: 420). The change in worldview to ecocentrism is unlikely to come from governments and business, the media, or even from the education system (Washington 2015). That leaves us – 'we the people'. A key aspect to changing our worldview is to rejuvenate our sense of wonder towards nature (see Chapter 9). When enough of us change our worldview, then governments, business and the education system will follow.

2) Reinventing ourselves to heal

This has been discussed in Chapter 7. It has been pointed out that our species is psychosocially undeveloped, that the path to healing the world will also require personal transformation and development, and a cultural or 'psychosocial' evolution within most societies (Hill 1991, 2014). This is possible, but we need to recognise such evolution is a key and needed part of the solution. As Chapters 3 and 5 have shown, most of us in Western society have grown up within a society that is fundamentally unsustainable, one based on an idea of endless growth, competition and inequality. Accordingly, most of us have (to a greater or lesser extent) been damaged within our inner selves. This is why we need to *reinvent ourselves* so as to help us heal.

3) Overpopulation

As Chapter 4 has shown, we cannot talk meaningfully about healing the world unless we talk about overpopulation. If we are going to accept reality, then we need to accept all of it. There are just too many of us humans living with a high (and increasing) consumption to be ecologically sustainable. Population growth is a key contributor to the environmental crisis. We need to stop denying this and develop rational, ethical, humane and non-coercive, ecologically based population policies (which exist, see Chapter 4). In the past, 'more people' made a certain sense in terms of growing food, building houses, etc. However, when the world is way beyond ecological limits more is no longer 'better', but far worse for both nature and humanity. Overpopulation *can* in fact be tackled by various non-coercive strategies (Engelman 2016). Denial of overpopulation keeps us from healing our world, as we remain in ecological overshoot beyond the limits of the gifts that nature can sustainably provide humanity.

4) Consumerism and the growth economy

This has been covered in Chapters 3 and 5. Is it 'uneconomic' to heal the world? This question is the wrong way around. We should instead ask whether it is economic *not* to heal it. It may cost a lot of money to solve our predicament, but it won't cost more than we already spend on the military (Pittock 2009). Brown (2011) estimates it will cost about an eighth of the world's military budget. Arguments of false economic expediency also need to be refuted. Failing to act on climate change and the environmental crisis (which may cause the collapse of society) would end up costing us *far more* than acting now (e.g. Stern 2006; Wijkman and Rockstrom 2012). The same applies to other environmental problems. Saving civilisation means restructuring the economy at 'wartime speed'. It also means restructuring taxes to get the market to tell the 'ecological truth' (Brown 2011). Our economy is broken and fundamentally unsustainable, but there is an alternative we can move to – a 'steady state economy' (Dietz and O'Neill 2013; Daly 2014). Chapter 5 showed that the consumer culture is fundamentally unsustainable. It does not make us happier or increase our true quality of life. Our throwaway society is literally trashing the Earth. It is overdue for us to come to grips with, and reject, this deliberately constructed (and fundamentally unsustainable) ideology of consumerism (Assadourian 2013).

5) Solving climate change

A key task of any meaningful concept of 'healing the world' is dealing with human-caused climate change – and its denial (Washington and Cook 2011). Human-caused climate change is a reality we must accept and act on (e.g. Pittock 2009; IPCC 2014). The overwhelming scientific consensus is that climate change *is* happening, and humans are responsible for increased global temperatures. But so what? Does it matter? Our civilisation evolved in the last 8,000 years in a period of stable climate, but we are now changing that radically. We face a major threat from climate change to the survival of our planet's ecosystems, and of course our civilisation relies on these. We face major impacts on our food supply due to rising temperatures and decreasing water supplies. We also face impacts on health, increased fire frequency, ecosystem change and expanding deserts (see Figure 1.2; IPCC 2014). We face social dislocation due to displacement of potentially hundreds of millions of climate refugees this century due to sea level rise (Dasgupta et al. 2007). So yes, *it does matter*. Climate change action is rightly seen as a key part of healing the world. It is not the only thing we need to do, but it is something we must urgently stop denying, so that we can carry out the solutions (which we have known for decades, Washington 2015). We need to accept that climate change is really a *symptom* of an unsustainable growth culture. We need to solve its root causes (overpopulation, overconsumption and the endless growth economy, Ibid.).

6) Keeping life alive

While today climate change is at last commonly discussed, we still rarely hear about the extinction crisis, which should worry us just as much (Rees 2017; IPBES 2019). The scale of extinction was discussed in Chapter 1 but must be highlighted again here to show the need for this essential solution framework. Because of our actions, a huge extinction event of life on Earth is underway, where a half (Wilson 2003) or even two-thirds of life (Raven et al. 2011) may be extinct by the end of the century. The recent IPBES (2019) report notes that one million species are currently threatened with extinction. The scale of such extinction is mind-boggling, rivalling the other five mass extinction events in the fossil record of the last 600 million years – except this time *we humans are responsible*. We are not talking about just a reduction in numbers of a few species, or some local extinctions, but the possible end of the evolutionary lines of half the life on Earth. It is like asking your child if she wants her children to be able to see an elephant *or* a tiger, as we cannot keep both (actually the outlook for both is worsening). As Soulé and Wilcox (1980) have observed: 'Death is one thing – an end to birth is something else'. We are in effect losing our own family (Crist 2012), as we are part of nature. The cause of extinction is due primarily to overpopulation (though even groups such as WWF in its Living Planet Report (2018) are coy about mentioning this), overconsumption and the endless growth economy. These are driving massive clearing of native vegetation, massive pollution and also accelerating climate change. Already many species are declining in range and abundance (Ceballos et al. 2015) and insect numbers have recently fallen by 75% over 27 years in protected areas in Germany (Hallmann et al. 2017) and by 40% around the world (Laurance 2019). As noted in Chapter 2, the ethical dimensions of what we are doing are huge. As Crist (2019) has noted: 'If anthropogenic mass extinction is left to run its course unchecked, how will the human psyche bear the onus of having massively extinguished our fellow earthlings?' Finally, if you read *The Overstory* by Richard Powers (2018), you will never feel about a tree quite the same way. So defend your trees, and if you can, plant a native tree to heal the world!

BOX 11.1 NATURE NEEDS HALF: VISION AND PRACTICE; SCIENCE AND ETHICS, VANCE G. MARTIN

Nature Needs Half (NNH) is both an aspirational vision and a common-sense, practical approach to living on an increasingly crowded planet. It is a real 'infrastructure' programme to protect and respect the wild nature that supports all life. Its goal is to ensure that enough wild areas of land and water are protected and interconnected to maintain nature's life-supporting systems and the diversity of life on Earth, to support human health and prosperity, and to secure a bountiful, beautiful legacy of resilient, wild nature.

NNH is science. Earth-systems experts agree on the correlation between the size of a natural area and its ability to maintain high-functioning biological processes that affect local and wider planetary systems. Depending on the type and location of the specific ecosystem or area of land or water, this threshold is roughly half of the original, pre-industrial extent. Below this threshold, nature struggles to produce critical eco-system services – in plain terms, 'life support'.

NNH is an ethic. Traditional cultures around the world are based on an ethic of reciprocity, of respecting and giving back to the Earth that supports us. In more formal religious teachings – such as Buddhism, The Society of Friends (Quakers), and others – a guiding principle is that of 'Right Relationship', through which one aligns personal behavior to honor all life.

NNH is also a metric to inform a much-needed, new model of human development. The good news is that NNH is as achievable as it is necessary – a global vision that can be achieved at all scales.

North America has numerous examples. One of the oldest is Yellowstone to Yukon (Y2Y), where a diverse team of conservationists, ranchers, planners, and wildlife lovers are creating a massive ecological corridor from the southern edge of the Yellowstone region in Northwest Wyoming to the Northern Yukon. Across Canada's Boreal north, one of the greatest forests in the world is the subject of both the highly destructive tar sands petroleum production operation of the Canadian government, and the location of massive conservation efforts by a wide range of players. First Nations are active in creating huge parks, such as the Nahanni National Park, expanded a few years ago to now be one of the 10 largest national parks in the world, at over 30,050 km² (11,602 sq. mi.).

Nature Needs Half can be applied at all scales, jurisdictions, and ownership regimes. Boulder County, northwest of Denver, Colorado (USA), is 62% protected; the Municipality of Hong Kong is 41% protected; the nation of Bhutan is 52% protected and connected; the Pacific Oceanscape is a consortium of 15 Island Nations protecting 40 million sq. kms of South Pacific oceans; and there are more.

Nothing less than bold vision and committed action at all levels is the formula to transform the juggernaut of exploitive human development. This will not be achieved by an organization, rather by a social movement that demands and directs the political will to do this. We can live with the Earth much better than we can simply live on it.

For more information: See: https://natureneedshalf.org/; www.half-earthproject.org/

7) Appropriate technology: a renewable future

This is discussed in Chapter 6. Techno-centrism, Cornucopianism and ecomodernism have been the hallmark ideologies of Western society, where the common idea is that technology can 'fix everything'. The environmental crisis clearly shows that it cannot actually do this. However, technology can actually help – if it's *appropriate*. One of the most frustrating aspects of our predicament is the denial of real and workable solutions using appropriate technology, such as renewable energy. We could move practically and economically to 100% renewable energy within two decades (if we are wise we will do this in just one). As a society, the science tells us we have to quickly cease burning fossil fuels. Fortuitously (just in the nick of time!) along comes a win/win technology – renewable energy. So why do many still deny this? Chapter 6 explains why and also discusses the various renewable technologies. For years most governments have buried their heads in the sand and denied the potential of renewable energy. However, the technologies exist, are feasible and affordable – all we need is the *political will*. However, we need also to abandon three very 'inappropriate' technologies: nuclear power, geo-engineering and 'carbon capture and storage' (Washington 2015). These are false solutions; in fact they are part of the denial of our key problems, and actually hinder us healing our world.

8) Reducing poverty and inequality

This was (in part) discussed in Chapter 1. As Rees (2010) notes, the world is in thrall to a grand socially constructed vision of global development and poverty alleviation centred on unlimited economic expansion, fuelled by open markets and more liberalised trade. However, rather than ending poverty, modernism has led to the environmental crisis, which is now impacting strongly on the poor (who rely on free ecosystem services, Washington 2015). The 'trickle-down' approach has proven to be cynical public relations. Inequality in income distribution has increased within OECD countries, and also in major emerging economies such as China and India (OECD 2011). However, poverty exacerbates environmental problems, which in turn exacerbate poverty (Washington 2015). An increasing population in poverty means they push ecosystem services beyond their sustainable level. Less available ecosystem services means the poor have less with which to live. We are in danger of going into a downward reinforcing spiral (MEA 2005).

Brown (2006, 2011) notes that in an increasingly integrated world, eradicating poverty and stabilising population have become 'national security issues'. Slowing population growth helps eradicate poverty (and its distressing symptoms), and conversely, eradicating poverty helps slow population growth (Brown 2006). With time running out, the urgency of doing both simultaneously is clear. A key way of reducing poverty (and consumer pressure) is improving equality of income (Wilkinson and Pickett 2010). One important solution is both minimum and maximum incomes (Dietz and O'Neill 2013). The developing world needs assistance to develop in smarter, more sustainable ways. Rich nations should spend 1–2% of their GDP on ecologically sustainable overseas aid to the developing world (in all our interests). After all, we live on the same planet in peril (Washington 2015). The developed world also needs to forgive much of the debt from old loans that the developing world still owes (and which legal contracts force them to keep repaying, holding them back from spending money on healing the key problems discussed here).

9) Education and communication

The importance of education has emerged in many chapters. The environmental crisis clearly represents a failure to properly educate and communicate. George Perkin Marsh wrote *Man and Nature* in 1864, yet he almost certainly had a more accurate view of humanity's role in the natural world than most decision-makers today (Daily 1997). Some 155 years later, most 'educated' people still remain unaware of his basic message: humans are part of nature and reliant upon it (Washington 2013). This fact is a condemnation of schools, universities and the media. It highlights the failure of professional scientists to adequately communicate this (Daily 1997). There is also the problem that education ideology can become an indoctrination that reflects the dominant cultural ideology. For example, Kopnina and Blewitt (2015) note that some critics have branded UNESCO's approach to environmental education as complicit in perpetuating the new 'holy grail' of the dominant political elites – namely the expansion of the consumerist culture. Such an approach will not solve an environmental crisis *created* by an endless growth ideology (Daly 2014; Washington 2015).

A key problem is that school curricula almost never cover ecology in the detail needed to explain human ecological dependence on nature (Washington 2018c). The problem extends even to universities, which Ehrlich and Ornstein (2010) observe are full of conservatism, entrenched interests and frequent sloppy thought. Academia is liable to get bogged down in unhelpful theory (see Washington 2018b), which does not in fact advance sustainability (and may hinder it, Washington 2018c). The issue is not just education; it is about 'empowerment' to take action. 'Facts' about environmental degradation abound; the trouble is finding how to *empower people* to take meaningful action. Hill (1998: 401) concluded: 'Indeed, the ability of a single, aware and empowered individual to bring about meaningful change should not be underestimated – assume you are that individual'. One essential aspect is to vastly expand 'education for wonder' (Washington 2018c, 2018d) in schools, universities and the community overall.

BOX 11.2 THE POLITICS OF HEALING: RESTORING ECO-LOGICAL AND CIVIC COMMONS, ERNEST J. YANARELLA

The politics of healing our broken planet, and increasingly fractured and polarized Western democracies, involves more than practicing the rights and duties of liberal citizenship. It also demands that an enlightened and activist mode of citizen action go beyond the activities of conventional politics – especially at a time when our ecological and civic commons are under threat by forces of corporate irresponsibility, political and economic self-interest, and narrow-minded and anti-scientific distortion. Nothing less is needed than the renewal of mass movements for political education, the mobilization of the courage to foster popular support for innovative strategies of resistance, and affirmative collective action. This requires the regeneration of hope in our democratic politics, and the concomitant refusal to give in to fatalism and despair.

In the realm of public policy we need to make changes, such as the rehabilitation of public understanding and trust in science, and a new willingness of politicians and unafraid citizens to imagine big and bold policies and programs – like the Green New Deal. Such policies need to equal the magnitude of the inchoate environmental problems on the horizon (unsustainable processes, 'global weirding', ecological scarcities). These policies must involve the scaling back of the bloated expenditures of national defense to reflect our true requirements for national security. In this restructured world order, our moral-political leadership will be founded upon collective security, international bodies, and the search for peaceful solutions to the underlying causes of terror, regional tensions, and other conflicts fueled in large measure by the U.S. and the West's contributions to the global arms bazaar.

In the process of expanding the realm of citizen action and social movements in politics, a worldview recognizing the entwinement of human beings within our more encompassing ecosystem must lead the way to the re-enchantment of nature as humans' 'organic and inorganic body'. This recuperation of nature as partner and creative force will lead to a remembrance that what we do to nature (deplete it and despoil it) we do to ourselves and our bodies (triggering physical and psychological dis-ease, and stunting our wonder and imagination). From the new sciences spawned by ecology, we are learning how multi- and inter-disciplinary perspectives on facets of nature all have more wondrous roles in the sustenance of life on our planet. This will lead prescient researchers to reconsider the body-nature-cosmos insights of religious and philosophical traditions of the East (Daoism, Buddhism, Confucianism).

The politics of healing then is not only grounded in new political ideas, new tactics and strategies for promoting change, and new technologies of power and mobilization of mass action. It incorporates the courage to transcend conventional ways of doing political organizing, engaging in aggregating social interests, and generating winning coalitions of voters. It strives to design public policies ambitious enough to fit the size of the environmental challenge we face. It calls for laying the groundwork for a transformative politics based upon a new worldview reweaving nature and human beings. This prompts us to raise our sights, free our creativity and imagination, and assemble those collective means to restore our ecological and civic commons – the true and only route to the politics of healing (Reid and Taylor 2010).

For more information:

Reid, H. G. and Taylor, B. (2010) *Recovering the Commons: Democracy, Place, and Global Justice.* Champaign-Urbana, IL: University of Illinois Press.

10) The politics of healing

Many of us shy away from politics, but this is really just a form of denial (Washington 2015). It is pure laziness, a way of shirking any responsibility to 'do something'. If we all do this then nothing will change and the living world (and society) will continue to decline. Disdain for politics is really apathy. Avoiding politics is thus failing to attempt to heal our world. Many politicians would like to act on the environmental crisis, but they need to know they have community support. That means *you* need to tell them. You need to be politically active about healing our world, you need to write letters to papers and local representatives. Your effectiveness and enjoyment of activism will be enhanced by joining an environment, sustainability or climate action group (Diesendorf 2009). We know change to heal our world is unlikely to come from the business community (though some have seen the light). Change is unlikely to come from the education system (but hopefully this will change, Washington 2018d). It is also unlikely to come from governments – *unless we make it happen.* We need to insist our leaders actually act on these ten healing strategies. Not just talk, but urgently act. As Monbiot (2019) observes:

> Those who govern the nation and shape public discourse cannot be trusted with the preservation of life on Earth. There is no benign authority preserving us from harm. No one is coming to save us. None of us can justifiably avoid the call to come together to save ourselves.

Critically, the politics of healing means society needs to abandon a politics of endless growth. Governments should no longer be elected purely based on promising a strong and growing (but fundamentally unsustainable) economy.

What we need is an ecologically sustainable (steady state) economy. Politics is at a crossroads; the choices are either affluence with persistent disparity, likely ecological collapse, or moderation with prospects for harmony between humanity and nature (Sachs 2013). If there is to be some kind of prosperity for *all* the world, the Western model needs to be superseded, making room for ways of living and consuming that leave a light footprint (Ibid.). Harich (2012: 36) points out that we have a problem in politics as most of us believe:

> [o]nce you find the truth and tell people the truth, they will adopt the truth because it's in their own interests. It's impeccable logic, so much so it's the core of modern activism.

However, he notes (Ibid.) that in fact this *doesn't work*. However he suggests a solution:

> Fortunately, there's at least one way out. It's the high leverage point of *general ability to detect political deception*. Currently this is low. If problem solvers can unite to raise it to a high level the race to the bottom will collapse, leaving the race to the top dominant. Politicians will then respond correctly to the truth about problems *and their root causes* because it will now be in their best interests.

Harich (2012) then argues that the first things governments would then do is to defang *Corporatis profitis* and turn them into *Corporatis publicus* by changing their goal to one that's sustainable. He concludes (Ibid.): 'The wheels of commerce, culture, and community will then be turning to the beat of the right drummer'.

So don't switch off and ignore politics. As enlightened economist John Stuart Mill (1867: 36) noted: 'Bad men need nothing more to compass their ends, than that good men should look on and do nothing'. The only way to heal the world is through *activism*. That means doing things, it means political action and political lobbying. The maxim for politicians is 'one letter equals a hundred votes'. So if they get 1,000 letters, it seriously worries them. Hence your contribution really *does* count. Solutions exist; we can heal the world; it is not impossible or hopeless. However, one of our major problems is lack of political will. It is no good just sitting back and blaming this on 'politicians'. We elect them, they are our representatives. If they are not taking action, it is because we are not telling them forcefully enough (Washington 2015). We also need to move to greater participatory (or deliberative) democracy (Elster 1998) but also to ecodemocracy (www.ecodemocracy.net).

Accordingly, we need to accept the political nature of the task of healing. Without mass political action, we will not be able to apply healing strategies. If we want to reach a sustainable future, then we need to be active in environmental politics *now*. I am not advocating a vote for any particular political party. Whatever party you vote for will need people who tell them about the environmental crisis and the need to heal the world. Political action is thus a key part of the 'Great Work'.

Conclusion

The need to heal our world should be *obvious*. Just travel to your nearest big city and look around you. Just consider the declining environmental indicators shown in Chapter 1. Clearly, the world is wounded – and we need to heal it. Yet society and governments still overwhelmingly deny we have wounded our living planet, or that there is any need for urgent change. Only we healers together can turn this around – that is you, me and all the many millions who love our world and fervently seek to heal it. Yet so often people see the problem and fail to look in the mirror to see the solution – *ourselves*. We are all called upon now to heal our world; the time for dithering and denial has past. Now is the time to urgently act for healing.

What can *I* do?

There are many things you can do to support the ten solution frameworks above (adapted from Washington 2015). These are:

1) **Talk about worldview and ethics**. Become a champion for changing society's worldview to ecocentrism (see Chapter 2). Talk about nature's 'intrinsic value', ecological ethics and the need for both social justice *and* ecojustice. Raise it with friends and family. Argue the need for change. You may find more allies than you thought.
2) **Reinvent yourself**. Think about how you can reinvent and rejuvenate yourself (see Chapter 7). What of your worldview and ethics? What of your empathy, wonder and compassion? What about your own denial, your own passivity and lack of action? How can you empower yourself to do more? What of how you educate and raise your children? What small (but meaningful) actions can you take to make the world a better place?
3) **Overpopulation**. Dare to talk about the double whammy of overpopulation and overconsumption on a *finite* planet, that the two together multiply to cause unsustainable impact (see Chapter 4). Consider having only one child yourself (McKibben 1999). Argue for an ecologically sustainable population target for your nation. Explain that we need action on both overpopulation and overconsumption.
4) **Consumerism and the growth economy**. Challenge the mantra of the endless growth economy and consumerism (see Chapter 5). Point out there is another way – a *steady state economy* (Daly 2014). Argue for tax and subsidy shifts; reporting on the Genuine Progress Indicator; and for minimum and maximum incomes. Argue for a truly 'honest market' that factors in all environmental and social costs. Rethink, reduce, reuse and recycle products.

5) **Solving climate change**. Demand urgent action to *mitigate* climate change by quickly reducing the greenhouse gases society has added to the atmosphere (not just action on 'adaptation' that governments prefer to discuss). That means we need to demand a carbon price (preferably a carbon tax). Push all levels of government to rapidly reduce their carbon footprint and support renewable energy. Join a climate action group, write letters, go and see your local political representative, and keep pushing your political representatives to act right *now*.

6) **Keeping life alive**. *Act to save the living world*. Lobby that all further clearing of native vegetation (e.g. forests, wetlands, savannah) must cease. This is not a draconian suggestion as society has already cleared so much of the habitat of our world's unique biodiversity. 'Keeping life alive' means extensive ecological restoration; it means greater connectivity between natural areas; it means curtailing huge developments and controlling urban sprawl. It also requires support for the 'Nature Needs Half' or 'Half Earth' vision (Locke 2014; Wilson 2016; Dinerstein et al. 2017) where half of terrestrial lands are reserved to protect life. Point out that nature deserves ecojustice (Washington et al. 2018). Support a 'Nature First' principle in regard to planning (Washington and Kopnina 2019).

7) **Appropriate technology**. Demand 100% renewable energy from your governments *within no longer than 20 years*, with immediate strong 'Mandatory Renewable Energy Targets' and 'Feed-in Tariffs' for all types of renewable energy (see Chapter 6). Oppose construction of any further fossil fuel power plants. Oppose also the dangerous mirage of nuclear energy. Install renewable energy yourself (e.g. solar hot water, solar photovoltaic panels). Get a sustainability audit of your home so you can cut energy and water use.

8) **Reducing poverty and inequality**. Support fair trade products. Lobby for 1–2% of your nation's GDP being spent on ecologically sustainable overseas aid to the developing world. Support the forgiving of much of the old debt owed to the developed world by developing nations. Argue for greater equality of income, and for minimum and maximum incomes in society (see Chapter 1).

9) **Education and communication**. Talk about the need to heal our world, about the need for all three strands of sustainability (ecological, social and economic). Condemn tokenism and 'greenwashing', and argue for serious action in your community. If you have children at school, speak to the school about discussing the environmental crisis and becoming more sustainable. If you are at university, promote these ideas to make your university a hub for sustainability and healing. Work to empower your local groups to take action. Support 'education for wonder' programmes (Washington 2018d).

10) **The politics of healing**. Be politically active (in whatever political party), arguing we must heal our world. Join an environment or climate action

group. Write letters, or even better, go and see your local member (and keep on doing this as persistence does produce results). Demand s/he supports real change to heal our world. In particular, demand your representative supports a strong climate change and renewable energy action plan *right now*. Support greater participatory democracy and the introduction of eco-democracy (www.ecodemocracy.net).

BOX 11.3 HEARING NATURE'S CHORUS: ECODEMOCRACY AS A MORAL NECESSITY AND A KEY SOLUTION TO THE ECO-CRISIS, JOE GRAY

Across the planet, each hour of every day, countless decisions are made by humans that have a bearing on the health and fate of non-humans, the majority constituents of the Earth community. That these decisions are, on the whole, made without serious regard for non-human interests is evidenced by the ongoing mass extinction event – the overwhelming tragedy of modern society. A once-vibrant ecosphere is being homogenized, and there is needless suffering on an appalling scale. From the perspective of all but the most short-termist celebrant of human supremacy, bad decisions clearly outweigh the good ones.

There is no shortage of headlines that highlight life's grim trajectory, yet humans are still failing to sufficiently internalize the needs of non-humans to act in a way that heeds the broad interests present in the ecosphere. In all political contexts, democracy as currently practised betrays a large majority of Earth's inhabitants.

The continuing growth of the aviation industry and big agriculture's rape of the ecosphere provide obvious example outcomes of morally delinquent decision-making. But even among better-meaning enterprises, such as ecological research, so rarely is explicit recognition given to the interests of non-humans. Furthermore, biological conservation itself is now tending towards serving human interests above those of the at-risk species it was conceived to serve.

What can be done? A major alternative to anthropocentric fairness-based decision-making is ecocentric democracy, or 'ecodemocracy', which considers the breadth of interests. As in traditional democracy, the decisions reached will not suit all parties, but since they will be made only after listening to the full chorus of voices, they will have honoured the moral standing of humans and non-humans alike. There is little doubt that such decisions would push the ecosphere back towards its formerly vibrant state.

In practical terms, representation of non-human interests can be achieved by, for instance, the appointment of human proxies. Dissenting political theorists might say that non-humans cannot be brought into the ring of stakeholders because they cannot understand fairness-based decision-making, and that the appointment of proxies is inappropriate because non-humans lack the ability to object to the way in which they are represented. But should not qualification for representation come from the possibility of being subjected to an unethical outcome, not the ability to understand ethics? And, countering the second challenge, how difficult would it be to appoint independent watchers to help ensure that the proxies are adequately representing their designated interests?

Ecodemocracy will only flourish with a concomitant shift in society towards an ecocentric value system. At the same time, it is one powerful means of helping effect this change.

How does the 'rights of nature' fit in? Enhancing the legal standing of non-humans will go some way to adequately representing their interests, but there is a need for a more fundamental shift in the way that society behaves. Non-human interests need to be represented not just through legal levers but in all political, administrative, and informal decision-making systems.

Further information: visit GENIE at https://ecodemocracy.net/.

References

Abram, D. (2010) *Becoming Animal: An Earthly Cosmology*, New York: Vintage Books.

Assadourian, E. (2010) 'The rise and fall of consumer cultures,' in *2010 State of the World: Transforming Cultures from Consumerism to Sustainability*, eds. L. Starke and L. Mastny, London: Earthscan, pp. 3–20.

Assadourian, E. (2013) 'Re-engineering cultures to create a sustainable civilization,' in *State of the World 2013: Is Sustainability Still Possible?* ed. L. Starke, Washington: Island Press, pp. 113–125.

Berry, T. (1999) *The Great Work: Our Way into the Future*, New York: Bell Tower.

Brown, L. (2006) *Plan B 2.0: Rescuing a Planet under Stress and a Civilization in Trouble*, New York: W.W. Norton and Company.

Brown, L. (2011) *World on the Edge: How to Prevent Environmental and Economic Collapse*, New York: W.W. Norton and Co.

Butler, C. (2002) *Postmodernism: A Very Short Introduction*, Oxford/New York: Oxford University Press.

Catton, W. (1982) *Overshoot: The Ecological Basis of Revolutionary Change*, Chicago: University of Illinois Press.

Ceballos, G., Ehrlich, A. and Ehrlich, P. (2015) *The Annihilation of Nature: Human Extinction of Birds and Mammals*, Baltimore: Johns Hopkins University Press.

Chawla, L. and Cushing, D. (2007) 'Education for strategic environmental behaviour,' *Environmental Education Research*, **13** (2): 437–452.

Collins, P. (2010) *Judgment Day: The Struggle for Life on Earth*, Sydney: UNSW Press.

Crist, E. (2012) 'Abundant earth and the population question,' in *Life on the Brink: Environmentalists Confront Overpopulation*, eds. P. Cafaro and E. Crist, Georgia: University of Georgia Press, pp. 141–151.

Crist, E. (2019) 'Let earth rebound! Conservation's new imperative,' in *Conservation: Integrating Social and Ecological Justice*, eds. H. Kopnina and H. Washington, New York: Springer, pp. 201–218.

Curry, P. (2011) *Ecological Ethics: An Introduction*, Second Edition. Cambridge: Polity Press.

Daily, G. (1997) *Natures Services: Societal Dependence on Natural Ecosystems*, Washington: Island Press.

Daly, H. (2014) *From Uneconomic Growth to the Steady State Economy*, Cheltenham: Edward Elgar.

Dasgupta, S., Laplante, B., Meisner, C., Wheeler, D. and Yan, J. (2007) *The Impact Of Sea Level Rise On Developing Countries: A Comparative Analysis*, World Bank, see: http://elibrary .worldbank.org/doi/book/10.1596/1813-9450-4136. (accessed 27 March 2019).

Diesendorf, M. (2009) *Climate Action: A Campaign Manual for Greenhouse Solutions*, Sydney: UNSW Press.

Dietz, R. and O'Neill, D. (2013) *Enough Is Enough: Building a Sustainable Economy Is a World of Finite Resources*, San Francisco: Berrett-Koehler Publishers.

Dinerstein, E., Olson, D., Joshi, A. et al (2017) 'An ecoregion-based approach to protecting half the terrestrial realm,' *BioScience*, **67**: 534–545, DOI: 10.1093/biosci/bix014.

Edwards, A. (2005) *The Sustainability Revolution: Portrait of a Paradigm Shift*, Canada: New Society Publishers.

Ehrlich, P. and Ehrlich, A. (1991) *Healing the Planet: Strategies for Resolving the Environmental Crisis*, New York: Addison-Wesley Publishing Company.

Ehrlich, P. and Ornstein, R. (2010) *Humanity on a Tightrope: Thoughts on Empathy, Family and Big Changes for a Viable Future*, New York: Rowman and Littlefield.

Elster, J. ed (1998) *Deliberative Democracy (cambridge Studies in the Theory of Democracy)*, Cambridge, UK: Cambridge University Press.

Engelman, R. (2016) 'Nine population strategies to stop short of 9 billion,' in *A Future beyond Growth: Towards A Steady State Economy*, eds. H. Washington and P. Twomey, New York: Routledge, pp. 32–42.

Gare, A. (1995) *Postmodernism and the Environmental Crisis*, London/New York: Routledge.

Gough, S. and Scott, W. (2007) *Higher Education and Sustainable Development: Paradox and Possibility*, Abingdon: Routledge.

Hallmann, C., Sorg, M., Jongejans, E. et al (2017) 'More than 75 percent decline over 27 years in total flying insect biomass in protected areas,' *PloSONE*, **12** (10): e0185809, DOI: https://doi.org/10.1371/journal.pone.0185809.

Harich, J. (2012) 'The dueling loops of the political powerplace', *Twink organization paper*, May 5th 2012, see: www.thwink.org/sustain/articles/005/DuelingLoops_Paper.htm. (accessed 19 February 2018).

Hill, S. B. (1991) 'Ecological and psychological prerequisites for the establishment of sustainable agricultural communities,' in *Agroecosystems and Ecological Settlements*, eds. L. Salomonsson, E. Nilsson and T. Jones, Uppsala: Swedish University of Agricultural Sciences, pp. 28–51.

Hill, S. B. (1998) 'Redesigning agroecosystems for environmental sustainability: A deep systems approach,' *Systems Research and Behavioural Science*, **15**: 391–402.

Hill, S. B. (2014) 'Considerations for enabling the ecological redesign of organic and conventional agriculture: A social ecology and psychosocial perspective,' in *Organic Farming, Prototype for Sustainable Agricultures*, eds. S. Bellon and S. Penvern, Dordrecht: Springer, pp. 401–422.

IPBES (2019) 'Nature's dangerous decline 'unprecedented' species extinction rates 'accelerating'', press release Intergovernmental Science-Policy Platform on Biodiversity and Ecosystem Services, see: www.ipbes.net/news/Media-Release-Global-Assessment.

IPCC. (2014) 'Summary for policymakers', Fifth Assessment Report, International Panel on Climate Change, see: http://ipcc-wg2.gov/AR5/images/uploads/WG2AR5_SPM_F INAL.pdf (accessed 16 March 2019)

Jickling, B. (2009) 'Environmental education research: to what ends?' *Environmental Education Research*, **15** (2): 209–216.

Kopnina, H. (2012) 'Education for sustainable development (ESD): The turn away from "environment" in environmental education?' *Environmental Education Research*, **18** (5): 699–717.

Kopnina, H. and Blewitt, J. (2015) *Sustainable Business: Key Issues*, London: Routledge.

Laurance, B. (2019) 'Climate change is killing off Earth's little creatures', *The Conversation* Feb 12, 2019, see: https://theconversation.com/climate-change-is-killing-off-earths-little-creatures-109719 (accessed 5 March 2019).

Locke, H. (2014) 'Nature needs half: A necessary and hopeful agenda for protected areas in North America and around the world,' *The George Wright Forum*, **31** (3): 359–371.

Macy, J. (2012) 'A wild love for the world', Joanna Macy interview by Krista Tippett, 'On Being', American Public Media, November 1st, 2012, see: https://onbeing.org/programs/joanna-macy-a-wild-love-for-the-world/ (accessed 14 March 2019).

Marsh, G. P. (1864) *Man and Nature; Or, Physical Geography as Modified by Human Action*, New York: Scribner.

McKibben, B. (1999) *Maybe One*, London: Anchor.

MEA (2005) *Living beyond Our Means: Natural Assets and Human Wellbeing, Statement from the Board, Millennium Ecosystem Assessment*, United Nations Environment Programme (UNEP), see: www.millenniumassessment.org/documents/document.429.aspx.pdf (accessed 19 Feb 2018).

Mill, J. S. (1867) Inaugural address delivered to the university of St. Andrews, Feb. 1st 1867, see: https://books.google.com.au/books?id=DFNAAAAcAAJ&pg=PA36&redir_esc=y#v=onepage&q&f=false (accessed 27 March 2019).

Monbiot, G. (2019) 'Only rebellion will prevent an ecological apocalypse', *The Guardian*, 15th April, see: www.theguardian.com/commentisfree/2019/apr/15/rebellion-prevent-ecological-apocalypse-civil-disobedience

Nabhan, G. and Trimble, S. (1994) *The Geography of Childhood: Why Children Need Wild Places*, Boston: Beacon Press.

OECD. (2011) 'Towards green growth, organisation for economic co-operation and development', see: www.oecd.org/greengrowth/48224539.pdf (accessed 10/7/14).

Pittock, A. B. (2009) *Climate Change: The Science, Impacts and Solutions*, Collingwood, Vic/London: CSIRO Publishing/Earthscan.

Powers, R. (2018) *The Overstory*, London: William Heinemann.

Raven, P., Chase, J. and Pires, J. (2011) 'Introduction to special issue on biodiversity,' *American Journal of Botany*, **98**: 333–335.

Rees, W. (2010) 'What's blocking sustainability? Human nature, cognition and denial', *Sustainability: Science, Practice and Policy*, **6** (2) (ejournal), see: http://sspp.proquest.com/archives/vol6iss2/1001-012.rees.html (accessed 21 March 2019)

Rees, W. (2017) 'What, me worry? Humans are blind to imminent environmental collapse', *The Tyee*, 16 Nov 2017, see: https://thetyee.ca/Opinion/2017/11/16/humans-blind-imminent-environmental-collapse/. (accessed 2 March 2019).

Rickinson, M. (2003) 'Reviewing research evidence in environmental education: Some methodological reflections and challenges,' *Environmental Education Research*, **9** (2): 257–271.

Rolston, H., III (2012) *A New Environmental Ethics: The Next Millennium for Life on Earth*, New York: Routledge.

Sachs, W. (2013) 'Development and decline,' in *State of the World 2013: Is Sustainability Still Possible?* ed. L. Starke, Washington: Island Press, p. 125.

Soskolne, C. ed (2008) *Sustaining Life on Earth: Environmental and Human Health through Global Governance*, New York: Lexington Books.

Soskolne, C., Westra, L., Kotze, L., Mackey, B., Rees, W. and Westra, R. (2008) 'Global and local contexts as evidence for concern,' in *Sustaining Life on Earth: Environmental and Human Health through Global Governance*, ed. C. Soskolne, New York: Lexington Books, pp. 1–8.

Soulé, M. E. and Wilcox, B. A. (1980) 'Conservation biology: its scope and its challenge,' in *Conservation Biology: An Evolutionary-Ecological Perspective*, eds. M. Soulé and B. Wilcox, Sunderland (MA): Sinauer, pp. 1–8.

Stern, N. (2006) *The Economics of Climate Change (stem Review)*, London: Cambridge University Press.

Wackernagel, M. and Rees, W. (1996) *Our Ecological Footprint: Reducing Human Impact of the Earth*, Gabriola Island, B.C., Canada: New Society Publishers.

Wals, A. (2007) *Social Learning: Towards a Sustainable World*, Wageningen: Wageningen Academic.

Wals, A. (2010) 'Between knowing what is right and knowing that is it wrong to tell others what is right: On relativism, uncertainty and democracy in environmental and sustainability education,' *Environmental Education Research*, **16** (1): 143–151.

Washington, H. (2013) *Human Dependence on Nature: How to Help Solve the Environmental Crisis*, London: Earthscan.

Washington, H. (2015) *Demystifying Sustainability: Towards Real Solutions*, London: Routledge.

Washington, H. (2018a) *Kinship Poems*. Lulu.com, see: www.lulu.com/shop/haydn-washington/kinship-poems/paperback/product-23580118.html. (accessed 27 March 2019).

Washington, H. (2018b) 'Harmony – not theory,' *The Ecological Citizen*, **1** (2): 203–210.

Washington, H. (2018c) *A Sense of Wonder Towards Nature: Healing the World through Belonging*, London: Routledge.

Washington, H. (2018d) 'Education for wonder,' *Education Sciences*, **8** (3): 125, DOI: 10.3390/educsci8030125.

Washington, H., Chapron, G., Kopnina, H., Curry, P., Gray, J. and Piccolo, J. (2018) 'Foregrounding ecojustice in conservation,' *Biological Conservation*, **228**: 367–374.

Washington, H. and Cook, J. (2011) *Climate Change Denial: Heads in the Sand*, London: Earthscan.

Washington, H. and Kopnina, H. (2019) 'Conclusion: A just world for life?' in *Conservation: Integrating Social and Ecological Justice*, eds. H. Kopnina and H. Washington, New York: Springer, pp. 219–228.

Washington, H., Taylor, B., Kopnina, H., Cryer, P. and Piccolo, J. (2017) 'Why ecocentrism is the key pathway to sustainability,' *The Ecological Citizen*, **1**: 35–41.

WCED. (1987) *Our Common Future*, World Commission on Environment and Development, London: Oxford University Press.

Wijkman, A. and Rockstrom, J. (2012) *Bankrupting Nature: Denying Our Planetary Boundaries*, London: Routledge.

Wilkinson, R. and Pickett, K. (2010) *The Spirit Level: Why Equality Is Better for Everyone*, London: Penguin Books.

Wilson, E. O. (2003) *The Future of Life*, New York: Vintage Books.

Wilson, E. O. (2016) *Half Earth: Our Planet's Fight for Life*, New York: Liveright.

WWF. (2018) *Living Planet Report - 2018: Aiming Higher*, eds. World Wide Fund for Nature, M. Grooten and R. E. A. Almond, Gland, Switzerland: WWF.

CONCLUSION

Healing our world

But at times
Unexpectedly …
Catch a glimpse
Of deeper meaning,
Be able to share
Each poignant instant
In the palimpsest of life.
So precious,
So joyfully given,
Unique, wondrous.
Momentarily I am privileged
To touch grace
Content, complete.

From 'Moment to Moment' (Washington 2018a)

Imagine a world where a species evolved to a level of consciousness where she could think about the wonder of being part of life. Where she also learned to communicate and discuss it with others in her group. Imagine if they revelled in being *part of nature*, and were constantly amazed by the beauty and diversity of the nonhuman persons around them – their kin. This would be a truly intelligent and wise species, one which cared for its home and lived by respect and reciprocity with the rest of nature. The arts of this species would celebrate the wonder of their home and its life. This wise species of course would be committed to maintaining their home in a healthy way. This species would 'take', but would also *give back* to the land, for its key obligation would be to protect the wonder of life. Its society would thus be based on nurturing and healing. Today, there is no such major civilisation on our planet. However, if our civilisation fails to wake up and heal our wounded world quite quickly, then I can only put my trust in what Leopold (1949) called the 'odyssey of evolution'. If our species remains fundamentally *un*sustainable, refusing to live in harmony with life, then (like many others) it will pass into the fossil record. My hope would then be that the Earth will evolve past the mass extinction humanity created, and then return beauty and abundance to the world. Eventually, hopefully a *truly* intelligent, sustainable and wise species might evolve?

Or ... humanity could wake up today and become a truly intelligent and wise species that heals the Earth *now*, before we pass on into extinction (along with much of our living kin). We do not have to look far to find how to do this, for such wisdom exists in the cultures of the First Peoples of the Earth. These First Peoples (even now after mostly being relentlessly colonised) remain willing to teach Western society (now globalised around the world) how to again become a wise society. They can help us to heal our world, to heal the environmental crisis. That could happen if Western society stopped shouting about how we must 'always grow' – and actually *listened*. Listened to nature, and the 'wisdom of the elders' who learned sustainability from the living land. As I love the best in humanity, this option remains *my deepest hope*.

This book arose out of a fervent desire to help heal our world. If you have read this far you may also share that desire? As an environmental scientist who has written a lot about ecological economics and ecological ethics, I sought to share various facts, ideas and solutions. However, I also wanted to share the wisdom I have learned through over 40 years of thinking (and feeling) about what it is that has wounded our living world, and how to heal it. To be honest, I have come to the conclusion over several decades that much of academia remains committed to ambiguity and waffle, if for no better reason than that is the best way now to survive in academia. However, there can be no doubt (unless one remains wedded to denial) that there is *great urgency* to healing our world. To adopt the pluralism approach to solutions (see Chapter 11) that argues that everyone's solutions are equally right, might fall in line with current theory in academia. However, it fails to actually solve our predicament, as society's time and energy would then be frittered away on thousands of different (and often contradictory) solutions. Hence the focus of this book is on the ten key solutions (Chapter 11) we need to act on to heal the world. This book also focuses on what you *personally* can do to help heal the world, for each of us is a strand in the web of healing. Acting singly we can of course heal, but acting together we can heal far more effectively.

Now this book is not primarily written for academics; rather it is written for the educated layperson who wants to understand the predicament we are in. Then the reader may seek to solve the environmental crisis and heal our beautiful but wounded world. Hence each chapter starts with a summary. Each chapter then goes into the detail to support the statements in the summary. This detail contains the facts, rational arguments *and* ethical arguments for those who want to explore them. However, unlike most academics, I have been an activist seeking to heal aspects of our wounded world since I was 18. Accordingly, what I believe is most important now is that we all *act* together to heal the world. There is no time now for academic word games, apathy or denial, or corporate fantasies about 'growing' our way out of a predicament caused by the fundamentally unsustainable ideology of endless growth.

Our world is wounded by human stupidity, denial of reality, greed and a lack of listening, wonder and caring. Now of course, you will be able to turn on the TV, and within a short time you will find someone who glibly assures you that things are 'not so bad', or even that they are 'really good' and exciting. You will also hear various hacks speak excitedly about how we have to grow our way out of our problems – using the same ideas that got us into this mess in the first place. In the end it will be *up to you* to decide if you want to be fooled or not. Do you want to ignore all environmental science, along with ecological ethics (Curry 2011; Rolston 2012) and believe a much nicer sounding lie – and thus do nothing to help to heal? Alternatively, do you want to accept the grim reality of our predicament and seek to positively heal our world? In the end you will have to decide if 'ignorance is bliss', or whether it is instead one of the most foolish things one can do?

Now make no mistake, unlike some authors, I will never say things are hopeless, or that it is 'too late' (see Chapter 10 on 'optimism, pessimism and realism'). The solutions needed to heal the world are known and quite feasible (if challenging). Indeed, these solutions are also positive and exciting. They also bring the balm of knowing that one is actually acting to heal the world. There will *never* be a point where we say that it is not worth continuing the Great Work of healing the world. How could there be? Every healing strategy we undertake can only make things better. That doesn't mean that we will not lose more of nature, as we almost certainly will. And for those of us who love life, the ongoing loss of the natural world brings a grief that will be with us always (as it is with me almost every day). However, working for the 'Great Work' of healing the world helps to temper that grief. We need to understand we are not attending a wake for a dying world; rather we are working at the bedside of a wounded Earth – and seeking to *heal* her. We seek to heal her because we love life, we love nature and we value the harmony of people living sustainably with the rest of the world, our kin. So there is no point in saying 'Oh my God, it is too late!' as that is just an excuse for apathy. It does nothing to heal anything. Of course, we will each have to also deal with the grief about what is happening (and what extra loss the future will bring). However, put that in perspective – as you work to heal the world – by doing all that you *can* (without burning out!).

You may in fact need to reinvent yourself (Chapter 7) to be able to work past such grief ('nature rituals' certainly help). However, if we have any sense of kinship with the rest of life (as well as with our fellow humans), we need to accept our grief and move past this to heal the world. In fact we need to listen to the land, celebrate the land, respect the land and acknowledge a 'duty of care' towards her. And most importantly we must *love* this land more and more as our task of healing progresses. As Macy (2012) rightly concludes, the world will not be healed without our ongoing love. It follows that an essential action each of us can do is to provide the gift of our love and care *to* the living world.

However, I am also a practical scientist as well as a nature poet who listens to the wild. There are very *practical* reasons for the solutions in this book, based on the best environmental science, ecological economics, ecocentrism and ecological ethics (Washington 2015, 2018b). We need the scientific solutions and the solutions from appropriate technology (e.g. renewables). However, we equally need the right worldview and ethics to *guide* those solutions. Science and ecological ethics need to work together (rather than see themselves as worlds apart), and they must be inspired by our rejuvenated sense of wonder towards life (Washington 2018b). These are all central to healing.

Perhaps the most critical thing is to *break the denial dam*. We need to get society to abandon its denial of both the huge problems that face us, along with the solutions to fix these problems. That means you need to *speak out* and politely (but firmly) talk about the need to heal our world. For if society can finally accept the predicament it is in, then that is when the best in humanity can come to the fore. That is when our co-intelligence (Atlee 2003) can come to bear on our worsening predicament, when we can use our creativity and imagination to find better solutions, to speed up the healing that is so desperately needed. Of course, this is the 'long emergency', so our children will also need to carry on that healing. And the only way we can overturn denial is to talk about the reality of the problems (the science is well and truly in, and the facts are readily available) and the positive solutions that exist. Running parallel with this is the urgent need to talk about the need to ditch anthropocentrism and the 'Masters of Nature' ideology, and adopt and promote to others an ecocentric worldview and ecological ethics that upholds the intrinsic value of life and our 'duty of care' towards nature.

CONCLUSION BOX: A RETURN TO INTEGRITY, KATHLEEN DEAN MOORE AND MICHAEL PAUL NELSON

Let us grant that people want others to think well of them, and more importantly, that we want to think well of ourselves. So either we do what we believe is right, or we don't, in which case we tell stories to convince ourselves that what might look wrong is in fact right. Self-justification is a human imperative: often we would rather feel physical pain than feel moral shame.
This is significant.

It helps explain efforts to 'normalize' behaviors that will shock our progeny. Such life-withering, self-destructive acts as reducing ancient forests to toilet paper, killing and selling the last of countless species, sickening children, poisoning fields of food, and – most significant for our time –

marketing fossil fuels at such a rate that the consequent global warming threatens the stability of life on earth – all these must become *business-as-usual*, how things are done by normal, honorable people.

And we need to convince ourselves not only that this is *how* things are done, but how they *ought* to be done. Economies invent, or (perhaps more accurately) empower, the ethics they need. A consequentialist ethic is particularly well-suited for the cynical *ad hoc* justifications required by extractive capitalism. An act is right, the consequentialist says, if the value gained (usually human happiness) outweighs the value lost, considering both the quality and quantity of the value. Mining out oil and salmon and forests is fine, then, if the value of the happiness gained from profit outweighs the value lost, which it will, assuming there is no inherent value in minerals or fish or trees. Poisoning workers is fine, if the value of the coal or strawberries or soybeans to investors and consumers outweighs the workers' suffering and medical costs. Altering the climate and imperiling the very existence of future generations can be justified, if the value of future happiness is discounted, compared to our own present satisfaction.

But this kind of reasoning cooks the books, inflating the gains, discounting the losses. It leads people into moral peril, yes. But it also leads many people to want some better measure of their self-worth, a measure that takes account of their deeply-held convictions, the moral squirming they feel when they hurt a child (even for the sake of greater good), or approve a war (even for stable oil supplies), or recklessly burn fossil fuels (even when it's the only way to get to work). It's an intolerable situation, doing what we know is wrong because some skewed calculus or craven politician tells us it's right and, anyway, we don't have any choice.

In this context, the venerable ideal of *moral integrity* is useful and healing. Integrity, from *integer*, wholeness: a matching between what one does and what one thinks is right. To act lovingly toward the Earth, because we love it. To live simply, because we don't believe in taking more than our fair share. To foreswear practices harmful to the future, because we're not the kind of people who wreck a place before they leave it. A life built on integrity doesn't require excuses or tortured justifications or shame. What it *does* require is courage, truth-telling, and self-respect.

For further information: www.earthisland.org/journal/index.php/articles/entry/why-we-wont-quit-the-climate-fight/

Moore, K. D. (2016) *Great Tide Rising: Towards Clarity and Moral Courage in a Time of Planetary Change*, Berkeley: Counterpoint.

Nelson, M. P. (2016) 'To a future without hope', in *Moral Ground: Ethical Action for a Planet in Peril*, K. D. Moore and M. P. Nelson, Eds, San Antonio, Texas: Trinity University Press. See: https://static1.squarespace.com/static/5c3e38fd4cde7af964606022/t/5c60d24b9140b743f6497ab1/1549849164314/Moral+Ground+Nelson.pdf.

In conclusion, we all need to recognise that the scope of healing our world is indeed challenging. However, consider these points. First, this healing is necessary and essential for the survival of this wondrous living world – but also for our own survival. Second, it is ethical, the *right* thing to do. Third, it gives purpose to our lives – a healing purpose. Such a purpose is something we should be proud to strive for. As Moore and Nelson (2013) conclude, what we must have is *moral integrity*. Finally, we should also realise that seeking to heal is perhaps *the* most meaningful thing any of us can do in our lives.

Join us!

References

Atlee, T. (2003) *The Tao of Democracy: Using Co-Intelligence to Create a World that Works for All*, Manitoba, Canada: The Writers Collective.

Curry, P. (2011) *Ecological Ethics: An Introduction*, Second Edition, Cambridge: Polity Press.

Leopold, A. (1949) *A Sand County Almanac*, New York: Ballantine Books.

Macy, J. (2012) 'A wild love for the world', Joanna Macy interview by Krista Tippett, 'On Being', American Public Media, November 1, 2012, see: https://onbeing.org/pro grams/joanna-macy-a-wild-love-for-the-world/ (accessed 14 February 2019).

Moore, K. D. and Nelson, M. P. (2013) 'Moving toward a global moral consensus on environmental action', in *State of the World 2013: Is Sustainability Still Possible?* ed. L. Starke, Washington: Island Press, pp. 225–233.

Rolston, H., III. (2012) *A New Environmental Ethics: The Next Millennium for Life on Earth*, New York: Routledge.

Washington, H. (2015) *Demystifying Sustainability: Towards Real Solutions*, London: Routledge.

Washington, H. (2018a) *Kinship Poems*. Lulu.com, see: www.lulu.com/shop/haydn-washington/kinship-poems/paperback/product-23580118.html (accessed 27 March 2019).

Washington, H. (2018b) *A Sense of Wonder Towards Nature: Healing the World through Belonging*, London: Routledge.

IDEAS FOR EDUCATIONAL EXERCISES

This book puts forward facts about the environmental crisis, but also talks about worldview and ethics. It argues the solutions for healing the world depend on acting, where we both accept facts and reality, but also change worldview and ethics. The book also seeks to remove the 'taboos' about discussing difficult issues such as overpopulation and the growth economy. We need to have a dialogue about these if we are to heal the world. The ideas below are included to assist discussion and dialogue about healing solutions. These exercises suit educational classes (high school years 11 and 12 and university) but also community groups such as book clubs.

1 Get students/group members to 'adopt a solution' (student classes, book groups, etc.)

Ask group members to review the list of 'What can *I* do?' solutions and pick one (separately or as a small group) to implement in their lives over a specific period (e.g. the semester). At the end of that time get the group to report on their experience, successes and failures.

2 Pick a box!

Get students/group members to select one of the 19 boxes from the book for discussion. Get group members to either pick a box (or allocate them via a 'lucky dip'). Get each member to research the box topic, looking at related material, and report back to the group. The group will discuss each box topic after each presentation. At the end of all presentations have a general discussion about solutions to heal the environmental crisis.

3 Is the environmental crisis real? (group discussion)

Ask the group to research whether the environmental crisis is *real*. Group members should go *beyond* the book's references and do *their own* research. This could be brought together in an essay, or through a group debate.

Some suggested internet sites for their research are:

- 'World Scientists' Warning to Humanity: A Second Notice'. Alliance of World Scientists (2017). This has been signed by over 15,000 scientists, http://scientistswarning.forestry.oregonstate.edu/sites/sw/files/Warning_arti cle_with_supp_11-13-17.pdf.
- Millennium Ecosystem Assessment 'Living Beyond Our Means' (28 page summary), www.millenniumassessment.org/documents/document.429.aspx. pdf (see also www.millenniumassessment.org/en/About.html#).
- UN (IPBES) extinction report summary (2019), see: https://games-cdn.washingtonpost.com/notes/prod/default/documents/1c572aca-fe3d-4b5f-94d9-1885bfee7e7a/note/7561fdf9-27ed-4fa0-845c-dbef31bb3b0c.pdf #page=1.
- Center for Biological Diversity (re extinction crisis), www.biologicaldiver sity.org/programs/biodiversity/elements_of_biodiversity/extinction_crisis/.
- Global Footprint Network, www.footprintnetwork.org/.
- Living Planet Report, WWF, www.wwf.org.uk/sites/default/files/2018-10/LPR2018_Full%20Report.pdf.
- Stockholm Resilience Center (summary re Planetary Boundaries), www. stockholmresilience.org/research/planetary-boundaries/planetary-boundar ies/about-the-research/the-nine-planetary-boundaries.html.

4 Worldview and ethics

Most institutions of our society don't discuss worldview, ethics, values and rights very much. Yet arguably they should if we are to heal the environmental crisis. Get your group to consider the following points as an aid for discussion:

- Does our view of the world influence how we act and live, how we organise society, how we organise our economy?
- Are ethics (doing what is right or wrong) important in science and technology? If they are, why? If not, why not?
- Does nature have *intrinsic value* and a right to exist for itself? Alternatively, is it just a 'thing' for human use that has no other value than to satisfy our desires?
- People talk a lot about 'rights', but *what do they mean*? Is each right absolute, or do we have to consider all rights relevant to an issue? For example, most of us assume people have rights (such as the UN Declaration of Human Rights), but is this just for humans living *now* – or should we consider the rights of future generations? Similarly, does nature have rights too? If she

does, then maybe we have to accept that there are many rights that interact and that all need to be considered?

• Get the group to consider two 'thought experiments' about value (Sylvan in Curry 2011: 12):

Experiment A Imagine that all living organisms have been destroyed by nuclear war and the last person is the only living being, who will die soon. This person has the ability to destroy all diamonds remaining in the world. *Will she be destroying anything of intrinsic value?*

Experiment B Consider the case after nuclear war where all sentient creatures are gone, while plants remain. The last person intends to cut down the last tree, which *could* recolonise the world. *Will she be destroying something of intrinsic value?*

Such experiments get people to consider what they truly value. How one responds depends on whether one has taken an *anthropocentric* or *ecocentric* attitude to nature. For anthropocentric people, killing the tree is not wrong as it has no value and doesn't benefit people. For ecocentrics, cutting the last tree would be *ethically wrong* as it has value in itself.

5 Population

Population is a difficult issue to discuss rationally as it is a highly polarised and conflicted issue. One suggestion to get people to abandon pre-prepared positions is by way of this thought experiment that groups can discuss.

Gowdy (2014: 40) invites us to do a thought experiment where (painlessly) humanity's population is returned to a few hundred million, and ecosystems are restored to their former healthy condition. Ask the group what would happen in the future if that society kept the ideology of endless growth in population, in accumulation and in expansion of our current society? Given we have created the environmental crisis in around 100 years, how long would it take for that society (if it committed to endless population growth) to get back to a similar situation as we have today?

Hold a discussion with the group as to 'Why talking about population is a taboo'. Ask the group to imagine ways to address population size, and hopefully influence it in the right direction for sustainability. This might include questions such as: 'Are population "control" and/or coercive approaches ever justified?' If not, what steps to slow or reverse population growth might be consistent with individual students' (or group members) own values? Push each student to imagine what they might consider here. Or do some students simply feel the issue is beyond addressing in any comfortable way? If so, why might that be?

6 The economics of enough

Show your group a photo of the Earth from space and ask them: 'Where is the economy?' Point out we cannot see the economy because it is *an idea*. An economy is how we organise things in our society. Show the group this video

The Economics of Enough by Dan O'Neill of the University of Leeds www
.youtube.com/watch?v=WIG33QtLRyA&t=28s.

Then ask the group some questions:

- Should society *serve* our economy – or should it be the other way around?
- On a finite planet, can society keep growing physically forever?
- Should our economy operate within ecological limits, or is it OK for the economy to destroy nature (and the gifts it provides humanity)?
- Must society always demand more and more, or should we seek to cultivate an approach of 'enough is enough'?

Ask the group for ideas on how to change from an endless growth economy to one that is ecologically sustainable.

7 Consumption and consumerism

Get your group to watch *The Story of Stuff* (https://storyofstuff.org/) and look at the Simplicity Institute http://simplicityinstitute.org/ and look through the conference results of 'Enough is Enough' https://steadystate.org/wp-content /uploads/EnoughIsEnough_FullReport.pdf. Then ask them to debate some questions:

- How much is enough? What do we need for a truly 'good' life, or can we never have enough?
- Is 'Shop till you drop!' a good thing as an ideology? Why or why not?
- Gandhi said 'Live simply so others may simply live'. Given that over a billion people on Earth don't get enough to eat each day, is this a valid point?
- On a finite planet where society has pushed beyond safe ecological limits, do we think society can keep on consuming more and more materials and energy?

Then ask the group if alternatively we should promote *thriftiness* and *sufficiency*, rather than always consuming more and more?

8 Denial

Read the group these three quotes:

> 'The best way to disrupt moral behaviour' notes political theorist C. Fred Alford 'is not to discuss it and not to discuss not discussing it'. 'Don't talk about ethical issues' he facetiously proposes 'and don't talk about our not talking about ethical issues'. As moral beings we cannot keep non-discussing 'undiscussables'. Breaking this insidious cycle of denial calls for an open discussion of the very phenomenon of undiscussability.
>
> *Zerubavel (2006)*

'Elephants' rarely go away just because we pretend not to notice them. Although 'everyone hopes that if we pretend not to acknowledge their existence, maybe ... they will go away' even the proverbial ostrich that sticks its head in the sand does not really make problems disappear by simply wishing them away. Fundamentally delusional, denial may help keep us unaware of unpleasant things around us but it cannot ever actually make them go away.

Zerubavel (2006)

It is as if each of us were hypnotised twice: firstly into accepting pseudo-reality as reality, and secondly into believing we were not hypnotised.

Laing (1971)

Then ask the group if they know someone who *denies* something, whether health, financial reality, climate change or the environmental crisis? Ask them if 'ignorance is bliss', that is, is it OK for us to deny reality? Why or why not?

9 Appropriate technology

What technology is truly appropriate to heal our world? Read these quotes to the group:

Renewable energy expert Lovins (2006: 6) has asked re fossil fuels and nuclear power when compared to renewables:

> Why keep on distorting markets and biasing choices to divert scarce resources from the winners to the loser – a far slower, costlier, harder, and riskier niche product – and paying a premium to incur its many problems?

The International Renewable Energy Agency (IRENA 2018) states:

> Electricity from renewables will soon be consistently cheaper than from fossil fuels. By 2020, all the power generation technologies that are now in commercial use will fall within the fossil fuel-fired cost range, with most at the lower end or even undercutting fossil fuels.

Ask the group to discuss 'Why society still seems wedded to fossil fuels and nuclear power'. Is the problem big business, governments, fear of change, or what? How can we overcome the barriers to moving swiftly to 100% renewable energy (and thus contribute greatly to solving climate change)?

10 Reinventing society

Several scholars (and this book) argue that we need to reinvent society *and* ourselves personally. It is one thing to say society 'should change', but *how* do we go about this? Get your group to list ideas for changing society, then hold a group discussion as to how effective these would be, and how one would go about carrying them out.

11 Representing nature through a nature ritual

In Chapter 7 the 'Council of All Beings' was explained, outlining the steps to carry out the 'Council of All Beings' (see also www.rainforestinfo.org.au/deep-eco/cabcont.htm). Get the group to appoint a guide to assist the group through the ritual (it may be you or another member who volunteers). Encourage all group members to really put themselves in the shoes of another 'being' that they choose to speak for, where they will represent that being in front of the Council. When the Council meets in a circle they may wish to discuss a topic such as 'What is wrong with the world, how can we heal this?'

If you think this might be a bit too radical for your group, an alternative would be to split the group into groups of four, where two are humans (one from a rich nation, one from a poor nation), one represents future human generations, while one represents nonhuman nature. Ask them to really think about *who or what* they are representing and speaking for. Ask the groups to discuss a theme such as 'Enough is enough, things have got to change' or 'What is wrong with the world, how can we heal this?' Circulate around the groups and ask questions if it seems a group has stalled. Get each group to report on their consensus, but also their differences.

12 Collapse or transformation?

Anthropologist Margaret Meade has noted that throughout history, small groups of determined people have changed society. She said: 'Never doubt that a small group of thoughtful, committed citizens can change the world. Indeed it is the only thing that ever has' (used with permission, InterculturalStudies 2010).

This book argues that society must change so that it uses fewer materials, less energy and has fewer people and far less impact on nature. This could be by a form of collapse, or by a conscious active transformation of society to become truly sustainable. *Ask the group which they would prefer.* Then ask the group which of the two they think is more likely? Ask them to justify their answer. Then ask the group if they understand what a 'self-fulfilling prophecy' is? Get the group to discuss the statement: 'If we don't attempt change, nothing positive will happen'. Ask the group for ideas about what active change could take place in their area to help to heal our problems.

13 Is it unacceptably 'political' to talk about the need to change society?

Show this video to your group by 15-year-old Greta Thunberg speaking at the COP24 Climate Summit www.youtube.com/watch?v=VFkQSGyeCWg.

Then point out that often people know 'something is wrong' with the world, but do not understand quite *how* to change this. This is largely because schools, universities and most institutions don't teach people how to create change (Maniates 2013). Creating change means being an activist, which means lobbying governments, corporations and the community to make real changes (given the urgency of our predicament). Ask the group whether they think we (collectively as society) should teach people 'how' to create change – or is this too political? Ask them if it is wrong to be active politically, or whether it is time for such action? Ask the group for ideas to change society to heal the environmental crisis. Finally ask them if they are doing any of these things and *will they do some of them in the future?*

14 Environmental autobiography – water and wonder

Ask the group to consider the role that water has played in their lives. Give them at least a week (time for reflection) to write a short essay or create a PowerPoint on a particular high point of their lived experience with a body of water – or with a 'special place' in nature (perhaps from their childhood?) which they hold in awe, wonder and respect. Get the group to read their essay or show their PowerPoint presentation to the group. Then discuss with the group the importance of such places and why they can make a big difference to our lives.

References

AWS. (2017) 'World scientists' warning to humanity: A second notice', *Alliance of World Scientists*, see: http://scientistswarning.forestry.oregonstate.edu/sites/sw/files/Warning_article_with_supp_11-13-17.pdf. (accessed 2 March 2019).

Curry, P. (2011) *Ecological Ethics: An Introduction*, Second Edition. Cambridge: Polity Press.

Gowdy, J. (2014) 'Governance, sustainability, and evolution', in *State of the World 2014: Governing for Sustainability*, ed. L. Mastny, Washington: Island Press, pp. 31–40.

IRENA. (2018) 'Renewable energy power costs in 2017', *International Renewable Energy Agency*, see: www.irena.org/-/media/Files/IRENA/Agency/Publication/2018/Jan/IRENA_2017_Power_Costs_2018.pdf (accessed 15 March 2019).

Laing, R. D. (1971) *The Politics of the Family*, New York: Penguin.

Lovins, A. (2006) *Quoted in WADE (2006) World Survey of Decentralised Energy 2006*, Washington: World Alliance for Decentralised Energy, see: www.networkforclimateaction.org.uk/toolkit/positive_alternatives/energy/WADE_-_World_Survey_of_Decentralized_Energy06.pdf (accessed 22 March 2019).

Maniates, M. (2013) 'Teaching for turbulence', in *State of the World 2013: Is Sustainability Still Possible?* ed. L. Starke, Washington: Island Press, pp. 255–268.

Zerubavel, E. (2006) *The Elephant in the Room: Silence and Denial in Everyday Life*, London: Oxford University Press.

INDEX

Abram, David 33, 40, 148, 155, 160–2, 189
academia: failure re solutions 34, 36, 69, 76,
   81, 189, 197, 209; jargon, use of 175;
   word games in 47, 210
activism: City Hall, fighting 2–3; in
   co-intelligence 36, 177, 211; cultural
   evolution needed 178, 192; defanging
   corporations 200; difference, making a 2,
   5; lack of action 3, 31, 201; Mill's three
   stages of 162; need for 200; seeds sown
   178; will we act? 3, 49, 99; see also Box
   1.1; Box 7.1; Box 9.1; Box 10.1; Box
   11.1; Box 11.2; Great Work, the
anthropocentrism: 'Anthropocentric
   Fallacy' 32; bio-economical thinking
   160; 'Critical Social Scientists' 34, 126;
   definition 32; ethics of 32–3; 'Greater
   Value' assumption 88; history of 33;
   human supremacy 29–30, 33, 100, 133,
   139, 165, 188, 203; human superiority
   33; humanism 33; impracticality of 34;
   insanity of 30, 40; insidious nature of
   33–4, 40; 'New Conservation' 34, 126;
   psychology of 35; as self-obsession 40;
   'Sole Value' assumption 88; valuation in
   32–7; see also hubris; intrinsic value of
   nature; worldview
apathy 3–4, 124, 127, 179, 199, 210
Assadourian, Erik xvi, 56, 63, 71–2, 94–6,
   99, 158, 178, 191, 193
assumptions: academic failure to consider
   19, 88, 91; bizarre 91; circular theory of
   production 89; of consumerism 88, 93–6;

denial of limits 89; dominance of
   Western 100, 110; environment as
   externality 91; ethics of 88; free market
   88; of hate 100; impact on society 88;
   list of key 88; 'Nature First' 101; of
   neoclassical economics 88; substitution of
   capital 91; summary 87; thermodynamics,
   ignoring 90; truth, biasing view of 93,
   100; unstated 31; what can I do? 101; see
   also worldview

baseline shift 174
Berry, Thomas 2, 38, 48, 62, 138, 145,
   155–6, 166, 177, 179, 182, 190–1
Brown, Lester 1, 13, 22, 24, 48, 52, 54, 72,
   81, 109, 111, 133, 165, 190, 193, 197

Caring for Country see ownership
Catton, William 8, 48, 53, 57, 69, 99, 124,
   165, 172, 175, 190
change, creating: see activism; Great Work,
   the; solutions
climate change: denial of 31, 49–50, 62, 89,
   218; human-caused 193; impacts of
   13–14, 49; rapidity of change 13;
   runaway 3, 14–15, 69, 173; solutions
   193, 202; as symptom 75, 194; tipping
   points 3, 173; what can I do? 202; see also
   Box 3.1
coal xxv, 54, 56, 92, 116, 202, 212
collapse: biosphere, saving 14; blocks to
   healing 29, 32, 56; catastrophe 170–4,
   181–2, 191; climate 173; crisis 171–2;